Contents

Genetics
The Mystery and the Promise

Francis Leone

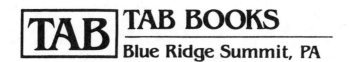

TAB | **TAB BOOKS**
Blue Ridge Summit, PA

12/92

FIRST EDITION
FIRST PRINTING

© 1992 by **TAB Books**.
TAB Books is a division of McGraw-Hill, Inc.

Library of Congress Cataloging-in-Publication Data

Leone, Francis.
    Genetics : the mystery and the promise / by Francis Leone.
       p.   cm.
    Includes index.
    ISBN 0-8306-3067-8 (p)
    1. Molecular genetics—Popular works.   I. Title.
    QH442.L46 1991
    574.87'328—dc20
                              91-20684
                                CIP

Acquisitions Editor: Roland S. Phelps
Book Editor: Stan Gibilisco
Production: Katherine G. Brown
Book Design: Jaclyn J. Boone
Cover Design and Illustration: Greg Schooley, Mars, PA

Introduction

Many popular books have endeavored to explain the essentials of modern genetics to the general reader and science-oriented student at the high school and college level. However, few such works give a detailed account of historical developments in the field and the methods by which the genetic code was deciphered. This book is intended to meet this task. The layperson interested in current knowledge of the gene, how it works, and how the concept came to dominate biological science will find much valuable information. The high school student who has completed a modern biology course should find this work excellent supplementary reading. The same applies to the college student of general science.

In the first chapters, much important historical material is presented, starting with Gregor Mendel's ingenious experiments with pea plants. These ground-breaking experiments prompted him to put forth the gene hypothesis. Some other subjects treated include Frederick Griffith's experiments with mice, the first real work showing that genes might be part of some chemical substance that we now know to be deoxyribonucleic acid, Avery's experiments that proved this conclusion, the Hershey-Chase experiment, the Menselson-Stahl experiment, the Beadle-Tatum experiment that initiated the one gene-one enzyme hypothesis, and Francis Crick's experiments with the T4 virus. The aim is to present such material in a concise, logical way that links each concept with others directly related to it. More than mere facts are presented. You will see how the gene went from being thought of as a segment of the chromosome to a portion of the DNA molecule and finally to a sequence of nucleotides in this molecule.

Just as extensive as the history is the account of modern chemical and physical methods that were brought to bear upon the breaking of the genetic code and the discovery of the chemical nature of the gene. X-ray diffraction was the first of these. Then came column chromatography, paper chromatography, and methods based upon the use of the ultracentrifuge. The chemical nature of the gene could not have been worked out without these techniques, each of which aided in finding a solution to the problem while they supplemented one another in the effort. You should find the text's treatment of these topics more information-packed than that of similar works. This presentation should bring the world of modern genetic research alive.

Finally, any book relating to contemporary genetics would not be complete without an account of biotechnology and genetic engineering, which have produced many new drugs and chemicals at lower cost than conventional methods. This area of genetic research is probably the most intriguing to lay people. You should find the information given in this area informative, stimulating, and thought provoking. No other field of genetics holds more promise for the future and peoples' well being.

Part |

Modern genetics and its beginnings

Chapter 1

Important historic developments

*G*regor Mendel was an Augustinian monk, a strange line of work for a scientist who made history. Yet he was a perfect example of a modern scientist. He not only unraveled some of the most basic laws of heredity and put forth a lasting theory to explain them, but also showed an ingenious use of the scientific method in his experiments. Perhaps his monastic education in physics and mathematics helped him in his notable quest, but many scientists have had the best of education without making any discoveries whatever. Genius and ability are often born in the individual; Mendel had both.

GREGOR MENDEL'S ENLIGHTENING PEAS

The times were ripe for Mendel's work. However, the world of science would not know about it until several decades had passed. Biology, the study of living organisms and their structure, had become a science, and the methods of science—observation and measurement—were being used in many realms with great success. Prior to Mendel's work, no precise experiments in the field of heredity had been done. However, some research had been performed in a way that left no room for definite conclusions about the laws governing the passing of traits from parent to offspring, while early investigators just compared parent and offspring in an overall, superficial way by looking at the offspring to see how it resembled the parents. In doing so, they were seeing many different traits showing themselves all at once. Such might bring to light a likeness between parent and offspring, although that is little more than you can observe in daily situations. Experiments are not necessary.

Mendel's incorporation of scientific genius and painstaking curiosity made him stand out from his forerunners in the effort to understand heredity and how it works. He realized something other scientists had not: The importance of looking at one single, easily observed trait at a time that could easily be controlled. You must look at a simple familiar trait and compare it in parent and offspring. An investigator could only get sidetracked by comparing organisms as a whole. To meet this simple requirement, Mendel chose a well-known organism for his experiments, the common pea plant, a neat choice because such plants have several easily seen traits that behave in a simple way.

Pea plants are either tall or dwarfed. The vines of dwarfed plants are less than a foot high; those of tall ones about three feet. Therefore, height is one simple trait exhibited by the plant that is well suited for study, since it shows itself in two ways, tallness and dwarfism. Seed (or pea) shape is another such trait. Peas in the pod are either round or wrinkled.

Mendel noticed other features of pea plants that made them ideal for use in his grand experiments. One of these features was that they are easily crossed, which aided in making many comparisons between parents and offspring because they could be grown in large numbers.

The overall experiment took much time, patience, and hard work and cannot be described in full here. The purposes of various parts of the investigation were much the same, so a few examples of Mendel's activities should serve to show how he made his findings. He had many seeds of what are called *pure* plants. Pure plants are varieties whose ancestors and offspring have the same traits as they do, no matter how many generations of plants are grown. Take pure tall and dwarfed plants to help make the concept of purity clear. If two pure tall plants are crossed and their seeds (or peas) are collected and planted, only tall plants will spring from these seeds, and, in turn, only tall plants will sprout from seeds obtained from this first generation of plants. Additionally, all third generation plants will be tall, and so on, while the two tall plants that started the ball rolling came from pure tall ancestors. The same behavior is found in pure tall dwarfed types. The two kinds of pea plants are said to be *pure* in the trait of height. In other words, a tall plant cannot sprout from a seed from two pure dwarfed plants, and a dwarfed one cannot arise from a seed from two pure tall ones.

In one part of his study, Mendel dealt with such pure tall and dwarfed plants that were the same in all traits except height. He grew a row of pure tall ones alongside a row of pure dwarfed ones and then crossed the tall ones in one row with the dwarfed in the other, his aim being to get seeds from them that would be crosses, or *hybrids*, of the original tall and dwarfed plants.

When the two rows of plants matured, he collected their seeds, planted them, and nurtured them with utmost care. What would this first generation of plants be like with regard to height, tall or dwarfed? Mendel waited for an answer in anticipation and anxiety. When the seeds gave plants that grew to maximum height and maturity, he found the plants were all tall; no dwarfed plants were among them, which was truly surprising. Had not one of the parents of each plant been dwarfed? Yet the trait seemed absent in all offspring.

Mendel devised another simple, but clever, experiment in which he crossed the tall first generation plants with each other. That is, he self-crossed them, and again collected their seeds and planted them. He wondered what height characteristics this second generation of plants would display. Dwarfism had not shown in the first generation. However, he wondered whether or not the trait had disappeared forever to be replaced by tallness in all generations of plants after the first. When his second generation plants grew, an astonishing fact came to light: Dwarfism had not ceased to exist. Quite a few second generation plants were dwarfed—about one out of every four. The rest were tall. Nevertheless, dwarfism had shown itself again. Although the parents of the second generation were tall, they must have had something in their internal makeup, reasoned Mendel, that gave rise to dwarfism in some of their offspring after it was passed into their seeds. Shortly, we will see what he thought this ''something'' was.

He then did similar experiments with pea plants differing in other traits, one bearing an experiment in which two kinds of plants, pure with regard to seed shape, were crossed. Plants having round seeds were crossed with others having wrinkled ones. Mendel found that seeds coming out of the cross gave plants having round seeds. Not one plant with wrinkled seeds was found in this first generation. When these round-seeded hybrid plants were self-crossed to give a second generation, both round and wrinkled seeded plants were found in that generation; again, about one out of every four plants had wrinkled seeds, which was the same kind of pattern Mendel had observed for plant height heredity.

He then paused and drew on his findings. What caught his attention was that one aspect of a given trait (for example, tallness or round seed shape) had totally displaced the other in all first generation plants, as when only tallness showed in that generation in his first experiment, while dwarfism did not show. Mendel called the first trait aspect that showed in his first generation hybrids *dominant*. The other aspect which did not show in them he called *recessive*. Tallness and round seed shape were dominant traits, while dwarfism and wrinkled seed shape were recessive traits.

What could account for these simple patterns in the heredity of peas? Mendel pondered this question deeply. Especially, why do recessive traits reappear in the second generations after being absent in the first? The theory he put forth to explain these happenings was quite daring in his day. It stated that traits like plant height and seed shape are regulated by tiny physical factors in the cells of the plant that were later called genes, which is what they are still called today. But knowledge about them has grown since the turn of the century. Each simple trait of the plant had two kinds of genes controlling it that formed a gene pair in each plant cell. There were three possibilities, for example, with the gene pair for plant height: the pair might hold one gene for tallness, represented by a capital letter like T, and one for dwarfism represented by d, or both genes of the pair may be for tallness, or both for dwarfism.

Let's look at how this scheme explained Mendel's findings. The pure tall plants he started with in his first experiments could be seen as having gene pairs for height made of two genes for tallness. These pairs could be given by the symbol TT, meaning each holds two genes for tallness. The pure dwarfed plants, on the other hand, had height gene pairs in their cells holding two genes for dwarfism. These pairs could be represented by the symbol dd.

On these assumptions, you can see why these pure tall and dwarfed types gave rise only to more of their own kind when self-crossed. When two pure tall plants are crossed, each contributes one gene for height to each seed, so that the height gene pairs in each seed has two genes (one from each parent) for the trait of tallness. Height gene pairs in the cells of the two parent plants each hold two tallness genes, or two T genes, so each plant gives a T gene to the height gene pair in each seed. The result is that the gene pairs for height in offspring sprouting from the seeds will hold two T genes, making them tall plants. The genetic composition of these gene pairs is TT and the offspring are pure with regard to tallness, because they only have height genes for that trait and none for dwarfism.

In the same way, two pure dwarfed plants have two dwarfism genes in their height gene pairs, which therefore have the genetic composition dd, and such parents can only give a d gene to each height gene pair of their seeds, making all their offspring pure dwarfed.

What did Mendel's theory have to say about the case in which he crossed pure dwarfed plants with pure tall ones, getting a first generation that were all tall? Again each pure tall parent could give only a T gene, while each dwarfed parent could give only a d gene to each of the seeds. Thus, each gene pair for height in the seeds held a T gene and a d gene and had the genetic makeup Td. However, T was what Mendel called a dominant gene, which would dominate the pair, while the dwarfism gene (d) had no control over height when paired with a T gene, even though it was present. That is why all first generation plants were tall.

Then came the second generation with both tall and dwarfed plants. How did the gene theory go over with this observation? Recall that this generation arose from a cross between tall parents having height gene pairs of the composition Td, which meant that the second generation plants could have height gene pairs of three different kinds. Each parent plant may give a T gene to a given seed, or one of them a T gene and the other a d gene to the seed, which will give a tall offspring plant in both cases because T is the dominant gene. A third possibility is that both parents give a d gene to the seed, which results in a dwarfed offspring. The second generation could therefore have three kinds of gene pairs for height: TT, Td, and dd; that, in turn, meant that both tall and dwarfed plants would be found among them, as indeed Mendel found that this was the case. The same kind of reasoning explained the heredity of seed shape in pea plants in exactly the same way.

Mendel's gene idea accounted for his findings with rigor, despite its controversial nature, and to him, at least, the heredity of peas was no longer a mystery. However, he had no clear picture in his mind about what his hereditary factors or units, later named *genes*, might be. Knowledge of that had to wait until chemists plunged into unraveling the chemistry of life. Mendel's work, by the way, was not even acknowledged by the scientific world for years after his death. But, at the turn of the century, three scientists—Hugo De Vries, a Dutch geneticist doing research on the heredity of evening primroses, Carl Corens of Germany, and Erich von Tschermak-Seysenegg of Austria—discovered parts of his original papers, read them, and found an attractive model for heredity in the gene.

Besides, the simple cases that Mendel studied are not the full story of inheritance by any means; the overall process is much more complex. Often, a given trait is controlled by many gene pairs, not merely by one, while in many instances a trait governed by one gene does not follow the simple patterns of dominance and recessiveness as did plant height and seed shape in these early experiments. Sometimes, neither of the two expressions of a trait are dominant or recessive. In certain plants, height gene pairs of the composition Td lead to a height shorter than that of the pure tall parent, but taller than that of the pure dwarfed parent. The offspring have a height between those of the parents. The T gene does not dominate the d gene, which, in turn, has no dominance over the pair either. Instead, the two genes blend to give a height between those of the parents.

Mendel's study of the simplest cases brought him insight that led him to postulate his gene theory, which still survives today and has come to dominate the entire study of heredity, showing how great a scientist the Austrian monk was. A skilled investigator always starts with the simplest cases and gradually works to the more complex. Mendel was the first real talented experimenter in the field. Tackling heredity as a whole, in all its diversity and complexity, was what had led previous pioneers astray.

What are genes? By what means do they regulate the many traits an individual organism has, be it a single cell or a fully developed human being? When chemists and biologists embarked on their exciting journey into the chemical composition of living matter, the answers finally started to come.

FREDERICK GRIFFITH'S MICE

Bacteria are small one-celled organisms that invade the blood stream of larger animals and sometimes cause disease in them. Strains of bacteria causing disease are called virulent, while those that do not are called avirulent. In 1928, the English scientist Frederick Griffith was engaged in experiments with a type of bacteria that goes by the technical name of pneumococcus; it came in a virulent strain that led to pneumonia in animals, and an avirulent one that was harmless. The virulent variety were large bacterial cells having a smooth carbohydrate coating over them. The avirulent cells had no covering and were smaller with rough membranes. Thus, the carbohydrate coating could serve to identify virulent pneumococci. It was known that both forms of the bacteria could be killed by intense heat, since immersing them in boiling water did away with them.

Griffith tried an ingenious experiment. He injected into mice three different solutions that contained these pneumonia causing bacteria. That injected into one group of mice contained live avirulent pneumococci, while that given to a second group held dead, smooth, virulent pneumococci that had been killed by heat. The next injection, given to a third group, was a mixture of dead, virulent pneumococci and live avirulent ones.

Griffith hit upon a strange discovery: The first group of mice given the rough pneumococci were not affected; there was nothing strange about that since the rough strain is not disease causing. Equally expected was that the second group having dead, virulent pneumococci in their blood stream stayed well and healthy. Why not? The virulent bacteria in their blood stream were dead. However, and this is the essence of Griffith's discovery, the third group given the mixture of live, avirulent pneumococci and dead, virulent ones developed pneumonia. How could this be? Dead virulent pneumococci did not produce pneumonia in the second group, so why should they in the third? The answer had to somehow involve the dead, virulent pneumococci. But how? It was unthinkable that live avirulent bacteria could restore the dead virulent ones to life, because life was known to arise only from preexisting life as is still known today, and no one has yet restored dead cells to life or created live ones from dead chemicals.

There was only one answer: Some substance was being transferred from the dead, virulent bacteria to the live avirulent ones that converted the avirulent type to the virulent strain, a process known as transformation, because an avirulent strain was changed or transformed into a virulent one. Other cases of transformation were uncovered, and before long, the number of cases became quite large.

But Griffith could not figure out the nature of the substance transferred. It was apparent it had to contain a gene for the formation of the carbohydrate coating of the virulent strain of peumococci. Did that mean that genes were part of the molecule of some organic substance? It was really too soon to tell. But advances were being made toward an answer. Griffith's work with transformation served as an impetus to get others doing research designed to reveal the physical nature of the gene.

CELL NUCLEI AND CHROMOSOMES

It was becoming clear that the nucleus directs the chemical activities of the cell which include the constant creation of enzymes that aid in the production of other organic substances (both protein and nonprotein) that determines the traits of the cell and the organism as a whole. Scientists pondered the question: How does the cell nucleus exert its control? An answer came eventually through the discovery of a material called chromatin, found in cell nuclei. Biologists became masters of using powerful light microscopes to look at various organelles of the cytoplasm, and when such instruments were turned to the nucleus, a complex material was found there that became brightly colored in the presence of certain dyes; the material was named *chromatin* because of these coloring attributes.

Chromatin underwent an intriguing change just before cell division. It shaped itself into little, long, rod-shaped bodies known as *chromosomes*.

Now it is necessary to say a word about cell division for the benefit of readers not familiar with the process. A baby develops from a single cell in the mother's womb by cell division. The first cell of the new person is formed by the union of one of the mother's egg cells and one of the father's sperm cells. This first cell divides in the mother's uterus into two cells which grow a little until each divides in two once again, giving four cells that grow a while and then divide in two; such continues until a full grown infant arises. Biologists watched cell division of all kinds under their microscopes and found it was also the way many bacterial cells reproduce and multiply. They were able to follow chromosomes as the cell separates in two, and found that just prior to cell division the chromatin shapes itself into chromosomes, half of which go to one side of the dividing cell and half to the other. Then, as the cell divides down the middle, each half goes into one of the newly formed daughter cells, both of which form a cytoplasm and nucleus. But this would leave each daughter cell with half the number of chromosomes (or with half as much chromatin) as the original cell. That did not make sense because daughter cells were known to be identical to the parent cell in all respects. Only one explanation suggested itself. Each chromosome must take advantage of the raw materials in the cell nucleus to build a duplicate of itself just before cell division, a process called replication. Because of replication, the two daughters cells have the same number of chromosomes as the parent cell, not half as many as would be expected if half the chromosomes of that cell are given to each daughter cell. After cell division, chromosomes once again merge together to become indistinguishable in the form of chromatin.

Scientists soon mounted enough evidence to show that Mendel's genes are located in the chromatin of cell nuclei. That meant that the secrets of heredity had to be there, too. Besides, it was learned that each individual organism had a certain fixed number of chromosomes in all its body cells. Human body cells have 46 chromosomes. Every cell in the human body (with the exception of egg or sperm cells of the reproductive organs) has that number of chromosomes. It was also shown that each parent contributes equally to the heredity of an offspring; a chromosomal explanation was found for this, too.

It may be illustrated by looking at human birth. What actually is involved in such birth? A mother has certain cells in the uterus called egg cells, which unlike normal body cells, contain only 23 chromosomes, half the normal number, 46. When man and woman mate, the man injects much smaller cells, sperm cells, into the woman's uterus, one of which unites with one of her egg cells to form the first cell of a new person. Sperm cells also hold 23 chromosomes or half the normal number. Therefore, when egg and sperm come together, the resulting cell, the first cell of the new individual, has the normal number of chromosomes, 46. Half of them came from the mother, and half from the father, so that each parent makes the same contribution to the child's heredity.

THE CHEMISTRY OF CHROMOSOMES

All evidence showed that genes were located in the chromatin of cell nuclei. That could mean only one thing: The chemical nature of genes could be illuminated if the composition of chromatin could be worked out. The problem of separating the substances in this complex material and studying them was long and tedious. But it was solved, and credit must go to many different chemists and biologists. Full details cannot be given now: only results. But later in this book, more details of this development will be presented.

The material of chromosomes belonged to another class of organic substances, the *nucleoproteins*. These substances had molecules with the same amino acid units as ordinary proteins as well as others quite different. Substances having molecules constructed only of these other units had been obtained from cell nuclei through various chemical treatments by the American biochemist of Russian descent, Pheobus A. T. Levine, in 1910. Earlier, in 1890, a German chemist Friederich Miescher isolated such a substance but did not determine its composition; he merely noted it was not carbohydrate, fat, or protein and named it *nuclein*, after the place from which he obtained it. Later he did analyze it and found it contained phosphorus, an element not found in pure proteins. Eventually Levine and others worked out the chemistry of the material. This will be discussed shortly. Nuclein became known as *nucleic acid*, and nucleoproteins were seen to be composed of both proteins and nucleic acid, although the protein units of a typical nucleoprotein molecule were thought to far outnumber the nucleic acid units.

This conclusion was based on the experience of biochemists of the time. All other biochemical substances that were mainly protein, having molecules holding other atomic groups, had been found to have protein contents that far outweighed the number of other groups present. Remember *hemoglobin*, the conjugated protein that carries oxygen to body cells. The same was assumed for nucleoproteins. Many biochemists came to think that the genetic substance (transferred from dead bacteria to live ones in transformation experiments) was mainly protein in nature.

In fact, early in the twentieth century, a theory of cell heredity based on protein arose. This was natural because it seemed logical that only a protein molecule could build something as complicated as another protein molecule. Nucleoproteins were seen as the genetic materials that directed the putting together of enzyme molecules that carried out the chemistry of both the cytoplasm and nucleus. However, the protein portion of the nucleoprotein molecule was thought responsible for the task. The nucleic acid part served some function less important. This view was destined to change.

It was known that chromosomes make replicas of themselves before cell division. Each chromosome takes chemicals present in the nucleus to make an exact copy of itself, thus doubling the number of chromosomes originally present. At the turn of the century, chromosomes were seen as mainly protein. It was therefore second nature to many biochemists and geneticists to picture chromosome replication as coming about because of proteins building other proteins like themselves. There were many enzymes in the average cell. If each of these alone, not to mention other structural proteins and others known as *hormones*, were to be a replica of some segment of some nucleoprotein molecule in the nucleus, many complex nucleoproteins would have to be present there. Each portion of their various chains would continually make copies of itself which would be one of the enzyme molecules the cell needed. This theory seemed both obvious and simple—perhaps too simple.

That is seen from the fact that evidence existed, back into the last century, that nucleic acids did the job; it was circumstantial evidence, but food for thought. One example concerned studies done on certain salmon sperm. What was astonishing about these sperm cells was that their nucleic acid content far outweighed their protein content, and worse, there was primarily only one kind of amino acid in the protein present—arginine. Others were present but were in the minority. Yet salmon cells had as many enzymes and other proteins as other cells on the average. Thus, the hereditary burden on their chromosomes had to be as great as on other cells, although, if the prevailing theory of the gene was correct, these salmon cells should have had no protein more complicated than this simple arginine-rich variety, which, of course, was not the case.

Despite mounting evidence that nucleic acid carried genes, many experts stuck with the protein concept with the hope that it could be modified to fit the facts. This was not blindness or stubbornness on their part. Most research at the time showed that nucleic acid molecules should be relatively small. They could not compare with those of proteins in versatility and complexity. It simply seemed self-evident to men of science near the turn of the century that the gene-carrying substance could only be a protein. There was no other alternative within the biochemical framework of the time.

No headway was made experimentally to settle the question either in favor of protein or nucleic acid until the substances that caused transformation were isolated in the test tube. Among the first pioneers in this area were three biochemists of The Rockefeller Institute: Oswald T. Avery, Colin W. MacLoed, and Maclyn McCarty. Their team did conclusive experiments in 1944 that showed the chemical nature of the genetic material beyond reasonable doubt.

GENES IN THE TEST TUBE

The substance given up by dead, virulent pneumococci held a specific gene for the making of the carbohydrate coating in live, avirulent pneumococci that absorbed the substance. The substance was most likely taken in by the avirulent cells through their membranes. But what was the substance? Griffith's indirect experiments with mice could not provide an answer. What was needed was a means to isolate the substance chemically, identify it, and show its effect on live, avirulent bacteria in the test tube; such was done by Avery and his team. Their entire experiments were rather complicated, although the main parts are easy to understand. More details will be given in the section on biochemical methods.

The first thing Avery did was simple: He took a solution containing dead, virulent pneumococci and separated the liquid from the dead bacterial cells. Using certain chemical procedures, he isolated the substances dissolved in the liquid and studied them; they were nucleoprotein in nature. He had not yet shown whether protein or nucleic acid accomplished transformation. Both were present among the dissolved substances. To determine this, Avery had to go further, and managed to separate the nucleic acid from the protein to get a solution of pure nucleic acid from the dead, virulent pneumococci. When live, avirulent pneumococci were allowed to grow and multiply in this solution, they absorbed the nucleic acid and became carbohydrate-coated, virulent pneumococci. Avery had shown that nucleic acid by itself could cause transformation. On the other hand, a solution of pure protein from dead, virulent cells did not cause transformation. Protein was not necessary, as least with the pneumococcus bacterial strain. Avery's findings stimulated many biochemists to do experiments with nucleic acids taken from many bacterial strains, and evidence was gathering in favor of nucleic acid as the gene carrier, though the fatal blow to the protein theory did not come until 1953; it had to do with large biomolecules known as *viruses*.

VIRUSES AND GENES

A virus is a large nucleoprotein molecule that breaks through the cell membrane and invades the cell. While inside it takes cell materials to make new viruses identical to itself through replication, thus interfering with the normal chemistry of the cell and sometimes killing it. A single virus may enter a live cell, but a hundred or more may emerge from the dead cell after invasion. All those that emerge are the same as the single virus that entered. By 1952, many viruses that infected bacterial cells were known. Two biochemists, A. D. Hershey and M. Chase, studied the infection process induced by viruses in cells and used bacterial cells going by the technical name of Escherichia Coli, or E coli for short, which are often found in the human intestine.

The main features of virus structure had been figured out; they were thought to consist of a central core of nucleic acid units surrounded by a shell of amino acid ones. Hershey and Chase wanted further proof that nucleic acids were the genetic materials. To find it they made use of the fact that the protein portion of a virus molecule contains sulfur atoms but no phosphorus atoms, while its nucleic acid core contains phosphorus atoms but no sulfur atoms.

Although atomic structure has not yet been discussed, the study of the smaller particles comprising atoms and how they are put together to make various atoms, the idea behind radioactive elements is easy to understand. Knowledge of it is needed to comprehend the Hershey-Chase experiment. Radioactive elements have atoms that are unstable and send out little, fast moving, energetic particles which are easily detected with certain instruments, one being the Gieger counter. Each time such a fast moving subatomic particle is detected by the counter, the counter makes a clicking sound. Radioactive sulfur gives out a different kind of particle than radioactive phosphorus and detectors can tell the difference between the two particles. Thus the scientist knows whether he is dealing with radioactive sulfur or radioactive phosphorus.

Hershey and Chase grew large colonies (or groups) of E coli bacterial cells in a liquid medium containing both radioactive sulfur and radioactive phosphorus. As the E coli grew in this radioactive liquid, they absorbed radioactive atoms of sulfur and phosphorus which became part of their biomolecules. At this point the experiments introduced viruses into the liquid which infected the cells and killed them. Viruses entering the radioactive E coli cells made many other viruses out of the cell materials, and, in doing so, took radioactive sulfur and phosphorus atoms and put them into new viruses as they formed and emerged from the dead bacterial cells. Although many nonradioactive viruses entered the E coli cells, many radioactive ones came out of them. In this way Hershey and Chase created many viruses with radioactive sulfur atoms in their protein shells and radioactive phosphorus atoms in their nucleic acid cores and separated them from the dead E coli cells for use in the second phase of the experiment.

The investigators then introduced these radioactive viruses into a medium containing normal E coli that was not radioactive while the cells were growing in it. The viruses entered the cells as usual and made them radioactive. After all the viruses did so, Hershey and Chase separated the dead cells from the liquid medium, examined the radioactivity of the dead E coli, and found it was caused by radioactive phosphorus. Essentially no radioactive sulfur was found in the dead cells. (A small amount of radioactive sulfur did find its way into the dead cells, though it was no more than could be accounted for in terms of inherent drawbacks in the experimental setup. So the experimenters felt they could ignore it.) Mainly radioactive phosphorus had found its way into the E coli cells during the infection process. But such phosphorus had been present only in the nucleic acid cores of the invading viruses, which meant that their cores alone had gotten into the cells.

When the two scientists studied the radioactivity of the liquid portion left after the dead cells had been separated out, they found it was that of radioactive sulfur, which had been situated in the protein shells of the viruses. That could only mean that the shells were left behind in the liquid when the nucleic acid cores entered the E coli cells.

WHAT THE RESULTS MEANT

The fact that only the nucleic acid cores invaded the cells showed that the ability to make new viruses lay solely with the nucleic acid of the virus, while the protein shell apparently served some secondary purpose. Not only new cores, but whole viruses like the original came out of the dead cells. So the nucleic acid portion not only directed its own reproduction, but that of the protein shell, besides. This showed beyond reasonable doubt that the ability of a virus to duplicate itself in a way similar to chromosomes is due to its nucleic acid portion only.

As other experiments were done it became clear that chromosome reproduction in the cell was directed by nucleic acid. Then what was the purpose of the protein in nucleoproteins? In the case of viruses it acts as an enzyme to speed up the process of boring open the cell membrane to let the nucleic acid core enter, while in the chromatin of cell nuclei it must also serve some secondary role, to be explored later.

Hershey and Chase helped show that nucleic acid alone has replicative ability; protein does not. So nucleic acid had to be the genetic material. But how? Remember it was thought that nucleic acid molecules were small alongside those of proteins. How could they ever be complicated enough to carry some kind of chemical instructions or code for the creation of all the complex protein chains vital to the cell? There was one sign of hope: bigger and better microscopes had been built. Many virus molecules are quite large—large enough to be seen with such powerful, efficient instruments. With them biologists were to observe individual viruses closely and could see their central nucleic acid part, which was seen to be as big as other parts of the virus and sometimes bigger, though the belief that nucleic acid was the minor part of chromatin still found ample support in Levine's original work. Which was telling the truth, chemical analysis or the pictures of large viruses seen under the microscope?

THE HISTORIC MESELSON-STAHL EXPERIMENT

No other molecule, natural or man-made, except DNA (and a few kinds of RNA) was known to be capable of creating an exact replica of itself. However, all evidence showed that such was the case with this large molecule. How else could biochemists explain replication of chromosomes seen under the microscope? These organelles are for the most part composed of DNA. And all aspects of a duplicated chromosome are normally the same as those of the mother chromosome that creates it from cell chemicals. It thus seemed to follow that chromosome duplication could only be

DNA-molecule duplication. But it was far from obvious how a DNA molecule of the chromosomes used free nucleotides found in the nucleus to make a copy of itself. The Watson-Crick model gave important clues, however, and led to a solution of this important problem of molecular genetics. The complementary nature of the base pairs of the DNA chains provided the key to understanding DNA-molecule duplication.

According to this model, the two chains begin to unwind at one end of the molecule just before cell division, and this exposes the bases on each chain one at a time; as each pair is broken during the unwinding, each of its dissociated bases attracts a nucleotide of the cell materials that holds its complement base, which becomes bound to it by means of the base pairing rules. The loosely attached base groups on the unwound portions of the chains are then joined to each other by means of a series of enzyme-controlled chemical reactions. In this easy way, two new polynucleotide chains are formed, that are complements of the originals. Each of the two new chains stays attached to the one in the original DNA molecule that guided its formation after the unwinding is completed. Two identical DNA molecules now exist where only one existed before. Each is normally the same as the original molecule. I say "normally" for reasons that will become clear later. Another molecule like the original has been created. But the important point is this: Each of the original chains or strands of the parent molecule acts as a guide for the putting together of another chain that is its complement and one of the chains in one of the daughter molecules.

How is it known that each of the chains of the parent molecule adheres to the one it creates in each of the daughter molecules? How could biochemists be sure that the two new chains do not pair up with each other? Such was not impossible. The new chains are complements of each other since they were formed by chains that are complementary in the original molecule. If such is the case, the new chains could somehow pair up with each other. Then both new chains would be situated in the new molecule. On the other hand, the two old chains would pair up once again to form the original molecule. However, if each new chain stayed with the old one that formed it, each molecule after duplication would hold one new chain and one old one. So the question before biochemists in the mid 1950s became: Does one chain of the original (or parent) DNA molecule exist in each of the two daughter molecules after duplication, or does one of the daughter molecules hold both of the original chains, while the other holds the two new ones? Clearly the first of these possibilities seemed the most reasonable and simple. Scientists prefer the simplest explanations of natural phenomena. But simplicity in itself proves nothing. Experiment had to decide by which of the two modes DNA replicates.

An ingenious experiment that showed that each new DNA chain stays attached to the old one that formed it was devised and performed by two biochemists, Meselson and Stahl, in 1958, and was based on the fact that all living organisms—even one-celled ones like bacteria and protozoa—

contain DNA in their cells. Large groups, or colonies as they are called, of bacterial cells can easily be grown and observed in a solution containing all the necessary nutrients. Bacterial cells reproduce by cell division. That made them ideal for DNA research of this nature. Those used in the Meselson-Stahl experiment were nurtured in a water solution in which the substance ammonium chloride, NH_4Cl, had been dissolved. Cl is the chemical symbol of chlorine, H stands for hydrogen and N is the symbol for nitrogen. As bacteria grew and multiplied in this solution, they absorbed ammonium chloride through their membranes. The ammonium chloride then passed into the fluids of their cytoplasm and eventually into their nuclei. There various enzyme governed reactions made purine and pyrimidine bases, using nitrogen atoms from the ammonium chloride units, NH_4Cl. These new base molecules ended up in the new DNA chains the bacterial cells made when dividing. So nitrogen atoms of the chloride in the nutrient solution, or growing medium, of the bacteria, ended up in the new DNA chains of the bacterial cells made during cell division. But if the experiment was to be of any value, Meselson and Stahl could not use only ordinary nitrogen in the experiment; they also had to use a heavier variety of the element. Ordinary nitrogen atoms are 14 times as heavy as the hydrogen atom. The heavier ones are 15 times as heavy. So ordinary nitrogen can be given by the symbol N−14, and heavier nitrogen by N−15.

In their historic experiment, the two biochemists first grew a large colony of bacteria in an ammonium chloride solution of N−15, the heavier nitrogen. All cells that grew in this medium therefore ended up with N−15 atoms in their DNA chains. In the second phase of the experiment, they took some of these bacterial cells and transferred them to an ammonium chloride solution containing ordinary nitrogen, N−14. Now, just before cell division, each chromosome makes a replica of itself. Therefore, as each bacterial cell divided in the ordinary nitrogen medium, normal nitrogen atoms went into the new DNA chains of the daughter cells, the first generation of cells after the first cell division. This second generation also grew and divided in the N−14 medium to yield a second generation of cells. These cells should have incorporated yet more N−14 atoms in their DNA chains. That much was clearly obvious. But the experimenters felt they could be somewhat more specific about the composition of the DNA in these first and second generation cells.

Because the first generation cells had been born in a solution of ordinary nitrogen atoms, their new DNA chains should have contained only normal nitrogen atoms—N−14—assuming of course that both chains did not break into pieces and reesemble themselves randomly during DNA replication, a possibility that was not absolutely ruled out at the time. On the same assumption, the original DNA chains, inherited from the parent cells grown in the N−15 medium, should hold N−15 atoms.

Now, this was the crux of the experiment: there were two possibilities for the kinds of DNA molecules that would be found among the first generation of bacterial cells, depending on how the DNA molecule replicates.

Assume for the moment that the newly formed chains remain attached to the old ones that guide their formation. On this assumption, we have one of these possibilities just mentioned. That is that each DNA molecule in the first generation bacteria should have one chain containing N−15 atoms, and one holding only N−14 atoms. On the other hand, what if the newly formed chains pair up with each other, while the old ones do the same? Then some DNA molecules in the first generation bacterial cells should contain only N−14 atoms in both chains. The rest would have N−15 atoms in both chains.

The two biochemists found that the first generation DNA molecules had only N−14 atoms in one chain and N−15 atoms in the other. How they determined this will be explained in the section on biochemical methods. This fact clearly indicated that the bacterial DNA replicates by a method in which the newly formed strands attach themselves to the old ones that formed them. This was one of the first decisive experiments that confirmed that DNA replicates by a method that puts one old chain and one new chain in each daughter molecule—a mechanism known as the *semiconservative* method of replication. If the two old chains stayed together while the new ones formed another DNA molecule, we would have what biochemists call the *conservative* method of DNA replication.

Meselson and Stahl then went further still in their experiment. They let the first generation cells divide in the ordinary nitrogen medium to give the second generation. As mentioned above, each of these second generation cells should contain some DNA molecules having only N−14 atoms in both chains. In fact, such DNA was found in them. This further showed that DNA duplicates semiconservatively.

What would have been found in the Meselson-Stahl experiment if DNA made more of itself by the conservative method? Then, the experimenters should have found only two kinds of DNA molecules in both generations of bacterial cells. One type would have both chains with N−15 atoms in them, while the other type would have two chains holding only N−14 atoms. Such was not the case.

The findings of the Meselson-Stahl experiment, along with other evidence from the Watson-Crick model, convinced biochemists that DNA replicates semiconservatively. Today this conclusion is accepted by all scientists.

Chapter **2**

Atoms, the chemistry of life, and cells

*L*ong before the chemicals comprising living organisms were isolated and studied, chemists knew they fell into certain broad classes sharing specific characteristics. Among them were three main catagories: carbohydrates, fats, and proteins. Proteins were the most important, being the builders of many parts of the living organism.

ATOMIC THEORY

Before pursuing these substances further, let's review some facts chemistry has revealed about matter, both living and nonliving. All material entities having weight and taking up space are made of matter consisting of many different substances, some common ones being water, sugar, table salt, wood, plastics, rubbers, iron, copper, tin, and mercury. But chemists discovered around 100 substances called elements that are basic; that is, all other substances are composed of two or more of them while they themselves are not made of any other substances but are the primary forms of matter out of which the thousands of other substances are constructed. Consider water, a substance well-known by all. By certain chemical techniques it can be separated, or broken down, into two other substances, hydrogen and oxygen, which are elements. Hydrogen is a very light gas, a form of matter like the air we breathe having no definite size or shape. A fixed amount of any gas will expand of its own accord to fill any container that holds it, and it also takes the container's shape, no matter what that might be. The light gaseous element hydrogen has no color, taste, or smell and burns vigorously in air when brought in contact with a flame. The burning is due to its chemically uniting with oxygen of the air to produce water, much heat, and light. Oxygen is a heavier gas that does not burn in air and is also colorless, odorless, and tasteless. Air is mainly a mixture of two gases, one of which is oxygen while the other, nitrogen, also an element, is a little lighter than oxygen and likewise has no color, taste, or smell. About one fifth of air is oxygen and four fifths nitrogen, with oxygen being the important, vital element humans and animals must breathe (or inhale) to sustain their life processes.

Hydrogen and oxygen are fundamental substances, or elements, and nothing chemists can do to them can change them into anything else the way water was changed into them. They are among the chemical building blocks of the universe.

As another example, take the familiar substance, table sugar. Intense heat (like that given off by the bottom of a pan placed over a hot burner of an electric range) will change white, sweet tasting sugar into dirty looking, black carbon and water. One substance, sugar, has changed into two others, carbon and water, on heating, with carbon being another chemical element which makes up a good part of the black ashes left after wood and coal is burned. But we have learned that water is composed of hydrogen and oxygen. So sugar must be made of three elements: carbon, hydrogen, and oxygen.

Table salt is a chemical combination of two elements, sodium and chlorine. Sodium is a silvery, soft, poisonous metal and chlorine is a greenish yellow gas with a choking odor that is also very poisonous. Eating a tiny piece of sodium or inhaling a small amount of chlorine can result in death.

So we have five elements thus far: hydrogen, oxygen, carbon, sodium, and chlorine. There are many others (a little more than one hundred) which include well-known substances such as iron, copper, silver, gold, mercury, lead, nickel, platinum, tin, aluminum, magnesium, phosphorous, and sulfur. All the elements are not equally common. In fact, only twenty of the lighter elements make up most matter around us while the rest are relatively rare with many occurring only in traces. The six elements, carbon, hydrogen, nitrogen, oxygen, sulfur, and phosphorus are among the most abundant and play a leading role in the chemistry of life.

Changes like water being separated into hydrogen and oxygen and sugar into carbon and water are examples of what chemists call *chemical reactions*, in which substances turn into other substances; they are very drastic changes. None of the characteristics of carbon or water can be seen in sugar. Carbon is a black, dirty looking, solid substance and water is a clear liquid that quenches thirst. Also, no one can see anything of hydrogen or oxygen in water or of sodium and chlorine in pleasant tasting salt, although chemistry reveals that two or more elements combined in the right way make a substance totally different from each of them. It is a fact of science that water and salt are made of these elements as is the observation that there are thousands upon thousands of such strange changes known to chemistry.

What could account for such total, bizarre changes in matter? Early chemists had ideas, but one that arose around 1803 seemed the best and was put forth by John Dalton, an English school teacher who did work in chemistry in his spare time. The theory has come to replace all others since it was proposed. Dalton postulated the modern atomic theory. His main point was that all matter must be composed of tiny, submicroscopic pieces or units called atoms, too small to be directly seen even with the most powerful light microscopes in existence today. Each element had its own kind of atom. Also atoms of one element differed from those of another in weight, size, and shape, and in how they combined with other atoms to form *molecules*. Since about 105 elements are known today, there are 105 different kinds of atoms that make up the world around us, and every object we see as well as the sun, moon, planets, and stars consist of atoms, the same kind of atoms known on earth. Chemists found a way to measure the weights of atoms. It turns out that hydrogen atoms are the lightest, while those of carbon are 12 times as heavy. Oxygen atoms are 16 times as heavy as those of hydrogen. To give you an idea of how small atoms really are, consider that one hundred million hydrogen atoms laid side by side form a line only an inch long.

Belief in atoms was not new with Dalton; they were part of the ideology of some ancient thinkers as far back as the fifth century B.C., one of whom was the Greek philosopher Democritus, although Democritus was not talking about a scientific theory and had his own ideas about atoms which he did not rigorously test through experimentation as the modern scientist would. Dalton's atomic theory, on the other hand, was a good scientific theory. The results of much observation and experiment showed it stood a good chance of being valid beyond reasonable doubt. Today the atom is seen as a scientific reality. No honest thinker could refuse to accept the idea.

But what did Dalton's theory have to say about chemical changes? Take the formation of water from hydrogen and oxygen to illustrate. In this chemical change, two atoms of hydrogen unite with one of oxygen to form a group of three atoms, the smallest possible particle of water known as a water molecule. In fact, all non-element substances consist of such groups of atoms called molecules; some elements also have molecules. The gaseous elements hydrogen, oxygen, nitrogen, and chlorine, for instance, have two-atom molecules made of two atoms of the element in question, while one variety of the element sulfur—the solid, yellow kind—has molecules composed of eight sulfur atoms bound to one another to form a ring. Molecules are the smallest particles of substances made of two or more elements, while the atom is the smallest possible particle of a chemical element.

Sugar molecules are a little more complicated than those of water or sulfur and consist of 45 atoms, 12 of which are carbon atoms, 22 hydrogen atoms, and 11 oxygen atoms. The nature of the forces that bind atoms together in this complex, and other more simple molecules, is beyond the scope of this discussion. There are two important points: elements are composed of atoms, and atoms of different elements come together in certain numbers to make the many diverse molecules that make up the millions of known substances.

However, chemists found that atoms obey certain rules in forming molecules, and behave as if they are capable of joining with other atoms in fixed numbers. For example, one oxygen atom joins with two hydrogen atoms to give a water molecule, no more, no less. Hydrogen molecules contain only two hydrogen atoms, and no more, implying that the hydrogen atom joins with only one atom of its own kind. Now consider the nitrogen atom. It joins with three hydrogen atoms to form the ammonia molecule. Ammonia is a colorless gas with a strong odor that is used in some laundry detergents. A carbon atom takes on four hydrogen atoms to form a molecule of methane, a flammable gas found in swamp or marsh gas.

It is as if each kind of atom had a number of hooks on it by which it attaches to hooks on other atoms. But no real hooks exist on atoms. All that is meant is that an atom seems to have a certain capacity to unite with hydrogen atoms. Chemists speak of *chemical bonds* on atoms, with the number of chemical bonds an atom has being defined as the fixed number of hydrogen atoms that atom joins with in forming a molecule made of that number of hydrogen atoms and itself. Thus, hydrogen atoms have one chemical bond, oxygen atoms two, nitrogen atoms three, and carbon atoms four. A chemical bond is represented by a small dash drawn out from the symbol of the element. In chemistry each element is given a one or two letter symbol. The symbol of hydrogen is H and that for oxygen O. Carbon has the symbol C, and nitrogen the symbol N. That of chlorine is Cl. In chemical bond notation the hydrogen atom appears as

$$H-$$

and the oxygen atom as

$$-O-$$

In the same way the nitrogen atom is given by

$$-\overset{\textstyle |}{\underset{\textstyle |}{N}}$$

and the carbon atom by

$$-\overset{\textstyle |}{\underset{\textstyle |}{C}}-$$

When molecules form, chemists say that the bonds on atoms connect. In the water molecule, where one of the two bonds on the oxygen atom joins to the single bond on one hydrogen atom, the other bond joins to that on the second hydrogen atom. The diagram of the water molecule in chemical bond pictures is usually given as

which is known as the *structural formula* of water. The structural formula shows the arrangement of atoms in a molecule. Likewise the structural formulas of ammonia and methane are:

```
        H                          H
        |                          |
      N H                      H-C-H
        |                          |
        H                          H
        |                          |
```

To understand what happens on an atomic and molecular level in chemical changes, like sugar changing into carbon and water on heating, and water separating into hydrogen and oxygen, let's look briefly at what modern science knows about the nature of heat.

Scientists discovered that atoms and molecules are in constant motion. Molecules of sugar, for instance, touch each other and are constantly vibrating about fixed positions. The intensity of this molecular motion decides the degree of hotness, or the temperature, of bodies; that is, the faster their molecules are vibrating, the hotter the body, or the higher its temperature. Therefore, as we heat a substance, we impart more motion to its molecules which begin to move faster. So as sugar is heated its big molecules start to vibrate about more quickly and bump against each other harder, which causes each of the large, relatively fragile molecules to break into 12 carbon atoms and smaller fragments, each made of hydrogen and oxygen atoms. The carbon atoms group together to form black carbon. Hydrogen and oxygen atoms in the fragments regroup themselves into water molecules. These adhere to each other, forming droplets of water.

This example shows that chemical changes involve the rearrangement of atoms into new molecules so that substances change into others; such apparently mysterious changes are thus well understood in modern atomic theory. Many complex chemical changes constantly take place in a single living cell, the smallest unit of living matter. Understanding of such changes through atomic theory has shed much light on the chemical machinery of life.

Matter comes in three physical forms or states: solids, liquids, and gases. Atomic ideas show why there is a difference among these forms of matter.

In gases, the attractive forces between molecules are very weak. They are so weak, in fact, that moving gas molecules stay apart and do not influence each other except when they collide. If more space is made available to them, they move into the additional space and the gas expands to fill the container holding it.

Attractive forces between molecules of a liquid are much stronger than those between gas molecules, and are strong enough to hold trillions of molecules together as a group in one area of space, although the volume (or size) of a fixed amount of liquid stays the same, since the molecules cannot break away from each other and wander freely like those of a gas. Water is a well-known liquid. A few ounces of water at the bottom of a drinking glass will take the shape of the lower part of the glass. But there will be a clear boundary—the surface—separating it from the space above.

Next comes the solid state. Most matter in the form of tables, chairs, rocks, and buildings is in the solid state. Solids keep a fixed size and shape when left to themselves. Rocks do not change in size or shape of their own accord, but must be hit with a sledge hammer or crushed with a large weight if they are to change form. The attractive forces between atoms or molecules in solids must be strong indeed: strong enough not only to hold the particles so close together that they touch, but also to prevent them from sliding over each other. As a consequence, solids are very rigid and do not flow. Their molecules only vibrate about fixed sites from which they cannot break loose.

Now it is a common experience that heating changes the state of many substances. Butter melts rapidly (that is, changes to a liquid) if heated in a pan on a hot burner of an electric range. Water can exist in all three physical states: as a solid, (ice), a liquid (such as tap water), or as a gas (steam). Heating a chunk of ice will eventually change it to steam. As the chunk is heated, its water molecules begin to vibrate about their fixed positions more rapidly, until finally a temperature is reached at which they are moving fast enough to break loose from one another's hold and slide over each other; the ice melts. As the water from the melted ice is heated further, it gets warmer as its molecules slip over one another faster. Eventually a temperature is reached at which they are moving fast enough to completely break loose from the attractive forces that hold them in contact. They then fly into space as free molecules. The water changes to steam, or boils, when that happens.

THE CHEMICALS OF LIFE AND CELLS

With this introduction to atoms and chemistry, we may turn to the complex substances of living matter. As mentioned earlier, the most important of these is a wide class known as proteins. Most molecules of living organisms are quite large, holding thousands, and often millions, of atoms, while many of these huge *biomolecules*, as they are sometimes called, are protein molecules. Since proteins are the builders of many parts of the living cell, the organism's diet must include many foods rich in proteins. Such foods supply it with enough of these essential substances for growth and maintenance of its internal chemical processes. Some foods rich in proteins are eggs, meat, cheese, bread, and fish.

Molecules of water and sugar are simple compared to those of most proteins. How did chemists figure out the exact details of the structure of such large molecules? Smaller molecules like those of water and ammomia had structures that were not too hard to decipher by chemical methods, since they hold only a very small number of atoms. But molecules holding thousands of atoms are another matter. The problem would have been insoluable if nature did not give chemists a needed break when they got around to these giant molecules. The key to figuring out the atomic makeup of such big molecules was this: Proteins and other giant molecules were found to be made of long chains of smaller molecules having structures and atomic makeups that were well-known or could easily be worked out. For proteins these smaller building block molecules were those 20 simpler substances called *amino acids* and the numbers, kinds, and arrangements of atoms in their molecules had been figured out quite well. The names and abbreviations of the 20 amino acids are shown along with their structural formulas in FIG. 2-1.

2-1 The amino acids.

2-1 Continued.

Lysine   Lys

```
            H   O
            |   ||
   H—N—C—C—O—H
        |
        H
        |
      H—C—H
        |
      H—C—H
        |
      H—C—H
        |
      H—C—H
        |
        N—H
        |
        H
```

Cysteine   Cys

```
            H   O
            |   ||
   H—N—C—C—O—H
        |
        H
        |
      H—C—H
        |
        S—H
```

Aspartic Acid   Asp

```
            H   O
            |   ||
   H—N—C—C—O—H
        |
        H
        |
      H—C—H
        |
      O=C—O—H
```

Phenylalanine   Phe

```
            H   O
            |   ||
   H—N—C—C—O—H
        |
        H
        |
      H—C—H
        |
        C
      /    \
  H—C        C—H
     |        |
  H—C        C—H
      \      /
        C
        |
        H
```

Methionine   Met

```
        H   O
        |   ‖
  H—N—C—C—O—H
        |
        H

      H—C—H
        |
      H—C—H
        |
        S
        |
      H—C—H
        |
        H
```

Proline   Pro

```
        H   O
        |   ‖
  H—N—C—C—O—H
        |
  H—C       C—H
     |         |
     H         H
        H—C—H
```

Serine   Ser

```
        H   O
        |   ‖
  H—N—C—C—O—H
        |
        H

      H—C—H
        |
        O—H
```

Valine   Val

```
        H   O
        |   ‖
  H—N—C—C—O—H
        |
        H

        H
        |
      H—C—C—H
        |
        H

      H—C—H
        |
        H
```

2-1 Continued.

Glutamine   Gln

```
              H   O
              |   ||
      H—N——C——C——O—H
          |
          H
              |
          H——C——H
              |
          H——C——H
              |
              C——N—H
              ||  |
              O   H
```

Glutamic Acid   Glu

```
              H   O
              |   ||
      H—N——C——C——O—H
          |
          H
              |
          H——C——H
              |
          H——C——H
              |
          H——C——O—H
              ||
              O
```

Alanine   Ala

```
              H   O
              |   ||
      H—N——C——C——O—H
          |
          H
              |
          H——C——H
              |
              H
```

Arginine   Arg

```
              H   O
              |   ||
      H—N——C——C——O—H
          |
          H
              |
          H——C——H
              |
          H——C——H
              |
          H——C——H
              |
              N—H
              |
              C=N—H
              |
              N—H
              |
              H
```

Asparagine   Asn

Threonine   Thr

Tyrosine   Tyr

Tryptophan   Trp

2-1   Continued.

Histidine   His

$$
\begin{array}{c}
\text{H}\quad\ \text{O} \\
|\qquad || \\
\text{H—N—C—C—O—H} \\
|\qquad| \\
\text{H}\qquad| \\
\\
\text{H—C—H} \\
| \\
\text{C—N} \\
||\qquad \diagdown \\
\qquad\qquad \text{C—H} \\
\diagup \\
\text{C—N} \\
|\qquad| \\
\text{H}\quad \text{H}
\end{array}
$$

Many protein molecules are not merely a single chain of amino acid units. Some consists of two, three, or sometimes four different chains connected to one another at certain places, while in others the chains are wound into loops or bend back on each other. The largest of these complex molecules have a structure in which the amino acid chains wind around one another in many varied ways. These are known as *globular proteins*.

It is now clear the characteristics of a given protein depend not only on the number and kinds of amino acid units in its molecular chains, but also on their arrangement in each chain, with each arrangement giving a different protein even though the same units may occur in the individual chains in each case. There are thousands of proteins in the average living organism. Nature has no trouble in supplying so many because of the versatile structure of the protein molecule. The many amino acid units in each chain can be ordered in numerous ways, giving a unique protein in each case.

The next important class of substance in organic matter are the carbohydrates, which are of two chief kinds, sugars and starches. Sugars have the simpler molecular structure of the two. Most sugars are white crystalline substances with a sweet taste; we have an example in table sugar used to sweeten our coffee. Another simple sugar is glucose. Instead of 45 atoms, its molecule has 24. Six are carbon atoms, 12 hydrogen atoms, and six oxygen atoms. Chemists use a shorthand notation to represent the atomic makeup of molecules, and in this notation the glucose molecule is written $C_6H_{12}O_6$, which is the chemical or molecular formula of this simple sugar. In this formula C is the chemical symbol of the element carbon and stands for one atom of the element. The little 6 at the lower right of the symbol says there are six carbon atoms in the glucose molecule. H is the symbol for hydrogen and stands for one atom of the element. Again the little 12 at its lower right means there are 12 hydrogen atoms in the glucose molecule. In the same way O is the symbol for oxygen, and we see there are six atoms of the element in the glucose molecule from its chemical formula. Water, a simpler molecule, has the formula H_2O, which says it contains two hydrogen atoms and one oxygen atom.

Combinations of element symbols like $C_6H_{12}O_6$ and H_2O are known as molecular formulas in chemistry and do not give as much information as structural formulas which show not only the numbers of various kinds of atoms in molecules but also give a good idea of their arrangement in space. For a simple substance like water, having small molecules, the molecular formula says as much as the structural one. There is only one arrangement for the three atoms in this simple molecule, and no others. And when we say a substance is simple we mean its molecule contains only a small number of atoms having one spatial arrangement.

But even simple sugars have molecules with enough atoms to give more than one spatial arrangement. Glucose is only one simple sugar with the molecular formula $C_6H_{12}O_6$. Two others with the same molecular formula are fructose and galactose, which differ from glucose and from each other because the atoms in their molecules are not arranged the same way, a phenomenon known as *isomerism*. Glucose, fructose, and galactose are said to be *isomers*, substances with the same molecular formula having different properties because the atoms in their molecules are arranged differently. Structural formulas of the three simple sugars are shown in FIG. 2-2. Most molecules in living matter are complex enough to have many isomers. We saw this in proteins, the chief molecules of living cells. Various arrangements of the same amino acid units lead to many different proteins, all having the same molecular formula. Calculations show that billions of different proteins can arise from one molecular formula. So many do not really occur in living tissues since the cell makes only a small fraction of those actually possible.

2-2 The structure of simple sugars.

Of the three simple sugars, glucose plays the central role in cell chemistry, though fructose or galactose may be used in its place if need be. We will see why shortly. For now it is enough to say that glucose supplies the cell with energy to carry out its chemical activities, though sometimes cells have a surplus (or excess) of glucose that cannot be used at the moment and a way is needed to store this glucose for future use. The cell does this by combining glucose molecules into a larger molecule, that of starch, which is made of many glucose units strung together. The type of starch depends on the number of glucose units comprising the large molecule and on certain features of their arrangement in the molecule. Foods high in starches are bread, potatoes, corn, lima beans, and pumpkin. Potatoes are composed of two kinds of starch.

Another group of energy producing organic substances are the fats. Fats are made of carbon, hydrogen, and oxygen like carbohydrates, although the number of oxygen atoms in the general fat molecule is much less than in carbohydrate molecules of the similar size. Fats are usually oily, or greasy, to the touch. Examples are many: corn oil, olive oil, cooking oil, solid fats on fresh meat, and that extra something that builds up around our waistline when we overeat. Fats are even greater energy supplier than carbohydrates. But the body mainly uses them as a means to store excess energy when the organism's food intake is greater than it can use at a given time, this being particularly true of humans. When we eat too much the extra food we cannot use at the time to drive our bodies, is converted to fatty substances.

STRUCTURE OF LIVING MATTER

Just as dead matter is composed of atoms, living tissue has been found to consist of small units called *cells*. Cells were first discovered by the English scientist Robert Hooke in the seventeenth century when he looked at a piece of cork under the microscope. The cork appeared granular. Each cork granule looked like a small empty compartment, and they were named cells. Actually the cork cells Hooke saw were dead, being only the shells of cells that had once been alive. Although most cells are too tiny to be seen with the naked eye, they are much larger than atoms, with the average cell containing billions of atoms arranged into many complex biomolecules. But these large molecules are organized in a unique way which gives rise to a state or condition called life. This state of matter is hard to define, even today. Whatever distinguishes living from nonliving matter still eludes chemists and biologists. On the other hand, much is known about the basic chemistry of cells and the chief characteristics possessed by living matter as opposed to dead, or inert, matter which include growth, locomotion (or the ability to move of its own accord), and, for plants, the ability to make food. When different kinds of cells group together to form different tissues of an advanced organism such as man, life takes on another attribute: a mind capable of thinking, reasoning, and planning. The study of the chemistry of life—*biochemistry*—still holds many mysteries, although much has been learned in the last thirty years in this exciting field. As atoms are the building blocks of dead chemicals, cells are the smallest units of living matter. There is no living creature smaller than a cell. The average cell is a tiny fraction of an inch in diameter.

In the human body, cells of one type make up one kind of tissue such as skin, muscle, or nerves, while certain tissues group up to make various organs like the eyes, tongue, brain, kidneys, intestines, and lungs.

A typical cell (FIG. 2-3), despite its small size, is a very complex structure. Biologists found that cells have tiny parts in them called *organelles* that play specific roles in the chemical activities that go on inside them. Most cells contain three main parts: a *membrane, cytoplasm,* and at least one *nucleus.* Usually there is only one nucleus, though there are exceptions.

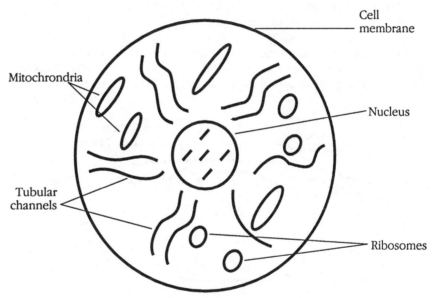

2-3 A typical cell.

The cell membrane is a thin film believed to be made of a layer of fatty material between two protein layers that encloses the contents of the cell. But enclosure is not its only purpose. It has small holes or pores over its outer surface through which nutritive liquids enter the cell. These liquids contain dissolved amino acids and glucose from the organism's food supply that are needed for protein building and energy production. The smooth part of the cell membrane regulates the passage of materials between the interior and exterior of the cell; it allows only certain types of molecules and not others.

A typical cell's interior consists of two parts: a small circular or oval organelle at or near the center called the nucleus, and a larger portion around the nucleus called the cytoplasm. The cytoplasm can be called the site of all the important, multi-complex chemical reactions taking place within the cell. These include the making of proteins out of amino acids and the slow burning of glucose to provide energy for chemical activities. The nucleus is the cell's control center, directing its own and the cytoplasm's activities, for the most part, although in recent studies it has been shown that there is a subtle interaction between the cytoplasm and nucleus in this regard.

The nucleus is surrounded by a thin membrane which separates it from the cytoplasm and has pores over its surface that let certain chemicals pass into the cytoplasm. One of these chemicals will be seen to be very central to heredity later.

The cytoplasm is a place of many different organelles, each playing a role in the chemical transactions of the cell, and is made up in part of water in which many substances such as amino acids and sugars are dissolved; it acts as a medium in which the cell organelles float. Some parts of the cytoplasm contain protein in the form of a gelatin-like medium (one having the texture of Jell-O) that holds other organelles and nutrient molecules. There are many types of cell organelles, each serving a specific purpose. Two very important ones are the mitochondria and ribosomes (FIG. 2-4). There are many of these two organelles throughout the cytoplasm.

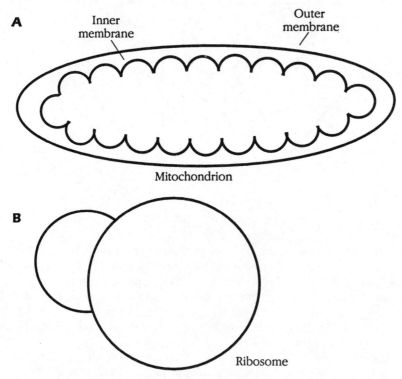

2-4   At A, a typical mitochondrion. At B, a typical ribosome consists of two parts: a smaller subunit and a larger unit.

Mitochondria are tiny, long, tube shaped structures with an inner and outer membrane of the same chemical composition, and have been called the powerhouses of the cell by many biochemists; in them carbohydrates such as glucose are slowly burned to yield energy to keep the cell running. When an individual mitochondrion is inspected closely under a microscope, little lump-like masses can be seen on its inner membrane which are believed to help in the burning of carbohydrates and fats. We will see how shortly.

Fats and carbohydrates burn in cells, but how does such burning differ from ordinary burning of wood and fossil fuels? Burning of wood, technically called combustion, is a chemical change or reaction. Wood, a substance made of carbon, hydrogen, and oxygen, has large molecules composed of atoms of these three elements. As a piece of wood is heated, its large molecules begin to vibrate about faster and bump into each other harder so that they break into fragments or groups of atoms. The carbon atoms of each fragment latch onto oxygen atoms of the air, forming carbon dioxide molecules (CO_2), while the hydrogen atoms of each do the same, giving water molecules (H_2O). Carbon dioxide is a colorless, odorless, gas heavier than air with molecules made of one carbon atom and two of oxygen as its formula shows. The process by which wood molecules combine with oxygen atoms yields much energy as heat and light. Everyone is aware that a bonfire is very hot and bright. But the burning of wood and fuels is fast. That of glucose in mitochondria is much slower.

Sugars, of course, are carbohydrates like wood, and while table sugar, for example, does burn when heated, it does so much more slowly than wood, so that we do not even notice it is burning. If we strike a match and put a sugar cube in the flame, the cube only seems to melt and turn brown. Most of the browning is due to sugar changing into carbon and water. And although a small portion of it may burn, the amount doing so is so small that it goes unnoticed. The same is true of glucose and fats, though they are the chief energy producers in cells. Why? These substances do not burn fast even at the high temperatures in a match flame. How can they do so at all at the low temperatures of the living cell? Temperatures in cells are a lot lower than those of a match flame, normal body temperature being only 98.6 degrees F—just slightly above room temperature. How could carbohydrate and fat burning occur at all under such circumstances? This was a serious stumbling block for biochemists trying to understand the chemical workings of life.

But another phenomenon was discovered that shed some light on the question. It can be illustrated as follows: Take a sugar cube and dip it into some finely ground black ashes—cigarette ashes will do—so that a thin layer of ash covers most of it, and then hold it in the flame. It then burns vigorously. The ashes have the effect of speeding up the burning of sugar, while an analysis of the charred remains of the cube would show they have not been changed in any way and are an example of what chemists call a catalyst, a substance whose presence speeds up a chemical change without being changed itself. Biochemists found that many chemical changes in cells are aided by protein catalysts called *enzymes*, which make it possible for glucose and fats to burn at the low temperatures found in the cell. The little lump-like masses on the inner membranes of mitochondria contain enzymes that speed up the burning to a pace at which it can yield sufficient energy to keep the cell going.

But combustion, fast or slow, involves free oxygen. So where do cells get such oxygen? All their oxygen is combined chemically with other elements in the large molecules of which they are built. The answer is simple: organisms breathe air holding free oxygen into their lungs, which pass the inhaled oxygen to the blood stream by means of the protein hemoglobin where it is distributed to body cells.

Hemoglobin is an example of a different type of protein than those mentioned previously; it is a conjugated protein, or one whose molecule contains a small number of atomic groups other than the usual amino acid ones. Proteins having only amino acid units are known as pure proteins in contrast to conjugated ones like hemoglobin. The hemoglobin molecule is composed of four connected peptide chains, each of which holds an atomic group called a heme group at one end. This is distinct from an amino acid group, since iron atoms are present in it and serve an important function in the molecule. Because of these iron atoms, free oxygen molecules weakly attach to the hemoglobin molecule. Now hemoglobin is situated in certain cells in the blood known as *red blood cells*. As blood flows through the lungs, the hemoglobin molecules in these cells latch onto free oxygen molecules taken into the lungs during breathing, and bind to them weakly so that these oxygen molecules are easily given up to body cells as blood circulates through the rest of the body. They are used by cells for the slow combustion of sugars and fats under the direction of *oxidizing enzymes* in the mitochondria.

Oxidizing enzymes hasten the combustion of sugars and fats. Combustion is a type of chemical change called *oxidation*, or the union of a substance with oxygen (or a similar element) to produce other substances—usually carbon dioxide and water, although such is not always the case. Oxidizing enzymes thus explain why burning can occur at the low temperatures in a living cell—a phenomenon coined ''the cold chemistry of living organisms'' by some early biologists and biochemists.

Enzymes have widespread importance in the cell. They are necessary to promote many thousands of chemical changes taking place therein. Enzymes are huge protein molecules (often globular proteins) that are quite complicated in shape with surfaces marred by many rugged extentions that make excellent disrupters of moving molecules that hit them. In other words, an enzyme works by providing a rough surface that helps break the chemical bonds between atoms in other molecules that take part in the reactions it speeds up. The process is known as *surface catalysis*; it is responsible for the effectiveness of enzymes. It is also the reason for another fact of biochemistry: a particular enzyme will hasten only one reaction in the cell and no others. That is, each biochemical reaction of the cell has its own enzyme and no other will do, which is to say that each enzyme has its specific purpose. No two different enzyme molecules have the same surface peculiarities; each has its own complex shape and no two have the same molecular architecture. Thus each enzyme can only break up, or activate, certain other molecules: only those that fit into the valleys and crevices on its surface. But enzymes are proteins. We have thus met another example of the protein's role as builder of living matter.

Where are enzymes and other important proteins made in the cell? The answer lies with another vital organelle, the *ribosome*. Ribosomes were found to be sites of protein manufacture and are oval in shape. They consist, for the most part, of an organic substance known as *ribonucleic acid*, or RNA for short. Ribosomes are the protein producing plants of the cell. Proteins are the most vital components of living matter, if for no other reason than that enzymes are proteins. The chemistry of life depends on enzymes. Cells could not function without them. Ribosomes are found mostly along certain membrane-like tubes that extend into the cytoplasm from the pores on the cell membrane. Such tubes carry solutions containing dissolved amino acids. These flow into the cell through the membrane pores and are necessary for protein creation, the chief purpose of ribosomes.

GENES, ENZYMES, AND CHEMICALS OF LIFE

In the scheme of life, genes produce enzymes, which produce the chemicals of life. Every enzyme is a protein. Every trait of an organism has results, in some way or another, from chemicals made through the organism's body chemistry. The making of these chemicals is controlled by one or more protein enzymes made in the ribosomes of some body cell, meaning that every trait depends on specific enzymes for its expression. Thus, enzymes carry out the process of heredity.

Such insight poses the question: What regulates protein synthesis (or the putting together of protein molecules from the simpler amino acid molecules) in the cell? This is essential to an understanding of the chemical basis of heredity. Enzymes have been shown to be the cause of all traits that organisms (from single cells to men) have.

The information given so far is from the standpoint of modern biochemical knowledge and was being gathered at the time the chemical nature of the gene was being worked out. Yet it has been given to provide an insight into the chemistry of life and cell structure. Such is necessary if one wants to comprehend recent historic events in genetics that shaped our knowledge of the behavior of genes.

Chapter **3**

The molecular structure of nucleic acids

*C*hemists early in the century knew much of the basic chemistry of nucleic acids as outlined here. But such as not enough to understand the gene and how it works. For one thing, DNA and RNA, though both nucleic acids, had different functions in the cell. As we have seen, DNA was found in the chromosomes of the nucleus and was suspected of being the gene carrier, as the work of Avery and others had shown. On the other hand, most RNA in the cell was found to be situated in the ribosomes of the cytoplasm, and was therefore thought to play a part in protein synthesis, although some RNA was also found in a small organelle in the nucleus, the *nucleolus*, of which more will be said later.

Besides, RNA and DNA turned out not to have the same kind of molecular structure. This is apparent from the way ribose and deoxyribose nucleotides are joined together in the two molecules.

NUCLEOTIDE ARRANGEMENTS IN NUCLEIC ACIDS

Biochemist Phoebus Levine's work with nucleic acids was a milestone in nucleic acid chemistry. But it had one unfortunate result. When he went about separating the DNA from the protein in cellular nucleoproteins, he used strong acidic and alkaline solutions to do so. It was not then realized that these strong solutions broke the long molecules into shorter chains of only four nucleotides apiece. So Levine and others concluded that they were justified in their belief that they had confirmed the then popular view that DNA and RNA had relatively small molecules that could not possibly serve as gene carriers. But further studies showed that nucleic acid molecules were indeed long, and composed of many nucleotides. The RNA turned out to be the simpler of the two kinds of molecules. It consists of ribose nucleotides (in most cases) in "the polynucleotide chain"—one holding many nucleotides.

From this point on, a capital letter symbol will be used for each atomic group which is the first letter of the name of the substance from which the group is derived. The symbols of the purines and pyrimidines are shown in TABLE 3-1. The phosphate group is represented by P, the ribose group by R, and the deoxyribose group by D. None of the names of the seven atomic groups begins with the same letter, so this notation should not lead to any confusion.

Table 3-1 Purines and Pyrimidines

Purines	Symbol	Pyrimidines	Symbol
adenine	A	thymine	T
guanine	G	uracil	U
		cytosine	C

In terms of these symbols the four ribose nucleotides become:

$$P—R—A$$
$$P—R—G$$
$$P—R—U$$
$$P—R—C$$

Likewise, the four deoxyribose nucleotides are:

$$P—D—A$$
$$P—D—G$$
$$P—D—T$$
$$P—D—C$$

Now, as an example, imagine an extremely simple RNA molecule of just two nucleotides, the general form of which is:

$$P—R—(Purine\ or\ pyrimidine)$$
$$P—R—(Purine\ or\ pyrimidine)$$

Purine and pyrimidine groups are called base groups, or bases, by biochemists. Any one of the four bases A, G, U, or C can be placed at the right end of each of the two nucelotides of this simple RNA molecule to give 16 different kinds of two-nucleotide RNAs, since there are four possibilities for each base group in each of the two nucleotides. The 16 different RNAs are written out explicitly in FIG. 3-1.

P—R—A | P—R—A	P—R—G | P—R—A	P—R—C | P—R—A	P—R—T | P—R—A
P—R—A | P—R—G	P—R—G | P—R—G	P—R—C | P—R—G	P—R—T | P—R—G
P—R—A | P—R—C	P—R—G | P—R—C	P—R—C | P—R—C	P—R—T | P—R—C
P—R—A | P—R—T	P—R—G | P—R—T	P—R—C | P—R—T	P—R—T | P—R—T

3-1   The 16 different RNAs.

A three-nucleotide RNA molecule has the general form

R—R—(Purine or pyrimidine)
P—R—(Purine or pyrimidine)
P—R—(Purine or pyrimidine)

and again, at the end of each of the three nucleotides any one of the four base groups can be placed, leading to 64 three-nucleotide RNAs.

At this point a pattern emerges from these simple examples that holds for any type of RNA. Consider a long polynucleotide chain of any number n of nucleotides. The pattern concerns the number of isomers this long RNA molecule can make. It can be stated as follows: the number of such isomers is equal to 4 multiplied by itself n times, or in algebraic notation 4^n. We can see that this is a gigantic number if n, the number of nucleotides, is large. The reader may well wonder whether or not some of the different base arrangements along the polynucleotide chain are the same structurally, although they first appear to be different. A simple case deals with the two-nucleotide chain in our previous example. Referring to that example, look at the simple RNA that contains the two base groups adenine and cytosine, the isomers of which are:

P—R—A   P—R—C
\| and   \|
P—R—C   P—R—A

In the first molecule, adenine precedes cytosine from top to bottom along the short chain. In the second molecule, cytosine precedes adenine. Now imagine that the second molecule is flipped over, so that its adenine group is now at the top and its cytosine group at the bottom. Would the resulting molecule be the same as the first? The answer is no.

First of all, it would be found that the ribose groups of each molecule would not have exactly the same atomic arrangement in three dimensional space, while the same would hold for the A and C groups. This may not be apparent at first. The two dimensional portrayal of molecular architecture does convey important details of how the two isomers differ. But it does not show three dimensional structures. Atoms have a certain arrangement in three dimensions for a given isomer of a given molecular formula, a feature that a two dimensional structural formula on paper cannot really bring out. So, for these reasons, the two simple RNA molecules above differ structurally. No means of putting the two molecules side by side in space could make them look exactly alike. They represent distinct isomers with the same molecular formula P_2R_2AC, in terms of the atomic groups P, R, A, and C, just as the simple sugars glucose and fructose are isomers with the same molecular formula $C_6H_{12}O_6$ in terms of the atoms C, H, and O. The same comments apply to any two RNA molecules with the same base groups in different orders along their polynucleotide chains. They are inherently different in molecular structure. The substances they represent do not behave the same chemically or physically.

Thus, any RNA molecule, and the RNA it comprises, is determined by three factors:

1. the number of nucleotides in its polynucleotide chain

2. the base groups on these nucleotides

3. the arrangement (or order) of the base groups along the chain

Changing any one of these factors will give a different kind of RNA.

THE STRUCTURE OF DNA

When biochemists came to determining the structure of DNA, matters were not as simple as for RNA. For one thing, DNA had been shown to be the gene carrier, while RNA played some secondary role. These facts tended to imply that DNA somehow had the more complicated molecule of the two nucleic acids. Let's look at these developments one at a time.

Since numerous experiments had established DNA as the genetic material, one problem presented itself to biochemists from the start: by what chemical means does DNA build proteins? While it had become clear in the 1940s that DNA molecules held deoxyribose nucleotides, such nucleotides had no resemblence to the amino acid units of proteins. Most biochemists of the time were still instilled with the idea that only protein molecules could make other protein molecules. They would act as model molecules in the nucleoproteins of the nucleus that build other protein molecules like themselves. Recall the strong hold this theory had on biochemists, geneticists, and biologists.

But more and more experimental evidence was implying that DNA was the genetic material. So, around the late 1940s, this protein theory of the gene began to fade out of life science. Maybe one molecule did not build another like itself in protein synthesis. Could it be that protein manufacturing in the cell was not that simple? Perhaps a molecule like DNA could somehow carry a chemical blueprint, or message, for the putting together of other kinds of molecules like those of proteins. As the twentieth century progressed, this new idea gradually won acceptance among life scientists. In other words, DNA in some way not then understood—that is, in the mid 1940s—contained a code that directed the assembly of all the varied and complex peptide chains of the living cell. The assembling was thought to be carried out by a series of chemical reactions taking place in the cytoplasm at or near the ribosomes. The research called on the talents, skills, and innovations of many great biochemists of all specialties and nationalities.

But what is the DNA blueprint? Only this molecule will reveal the workings and constitution of the gene.

Levine's original belief that the longest nucleic acid chains held at most four nucleotides had been dispelled. DNAs were found that held thousands of nucleotides. However, the four-nucleotide theory did not die overnight; it was modified, so to speak, with many biochemists holding the view that long nucleotide chains were made of a series of groups of four bases that repeat over and over again down each chain as shown in FIG. 3-2. The four repeating bases could come in any order in each group. But one arrangement would repeat again and again down the chain. Problems existed in this scheme too. On the basis of this theory, there would just not be enough variety among the DNA molecules of the nucleus to carry the many instructions required for the creation of the wide gamut of peptide chains found in cells of the average organism. The DNA blueprint somehow had to reside with the base arrangements along the DNA polynucleotide chain. Only these are capable of any variation, the sugar-phosphate groups being the same all along the chain. But all calculations showed that the four-nucleotide theory was far too simple to account for all the proteins found in the average cell. What was the solution to this puzzle? Could biochemists deny the wealth of experimental evidence that implied DNA was the gene carrier and continue to endorse the protein theory? No, as scientists, they could not. Either the four-nucleotide theory was wrong or their senses were deceiving them. But the theory would have one simple consequence that could be checked by experiment if it were true. Any DNA would have equal amounts of the four bases A, G, T, and C when chemically analyzed. This would be true because the numbers of each base group in any DNA molecule would be the same. One fourth of the total number would be A groups, one fourth of them G groups, one fourth T groups, and one fourth would be C

```
 ⎧ P—D—A        ⎧ P—D—G
 ⎪   |          ⎪   |
 ⎪ P—D—G        ⎪ P—D—T
 ⎪   |          ⎪   |
 ⎨ P—D—T        ⎨ P—D—A
 ⎪   |          ⎪   |
 ⎩ P—D—C        ⎩ P—D—C
 ⎧ P—D—A        ⎧ P—D—G      3-2  DNA chains in the four-nucleotide theory.
 ⎪   |          ⎪   |
 ⎪ P—D—G        ⎪ P—D—T
 ⎨   |          ⎨   |
 ⎪ P—D—T        ⎪ P—D—A
 ⎪   |          ⎪   |
 ⎩ P—D—C        ⎩ P—D—C
     |              |
 and so forth
```

groups. That would be true no matter what order the four groups had in each repeating group of four.

Chemists set out to find the number of base groups in various DNA molecules by experiment. They had to break a given DNA, which could only be obtained from cells in extremely small amounts, into its four free nucleotide substances. The job was not easy. They then had to figure out the small amounts of each of the four free nucleotide substances in a given DNA. There was one catch, however. The four deoxyribose nucleotides have different weights like different kinds of atoms do. Thus, a given weight of a given DNA could not be expected to give equal weights of the four nucleotide substances on analysis, even if its molecule held equal numbers of the four base units. Equal numbers of the deoxyribose nucleotides would not weigh the same in grams, since each of them has a different weight.

Because of situations like this, chemists have adopted a special kind of weight measure. Take hydrogen and oxygen atoms to illustrate. It was found that the oxygen atom is 16 times as heavy as the hydrogen atom, and chemists calculated that there is an incredibly large number of tiny hydrogen atoms in just one gram of the element—about 602,300,000,000,000,000,000,000. Since the oxygen atom weighs 16 times as much as the hydrogen atom, that same number of oxygen atoms would weigh 16 grams. Now the carbon atom weighs 12 times as much as that of hydrogen. Thus, that number of carbon atoms would weigh 12 grams. A water molecule, having two hydrogen atoms and one oxygen atom, is 1 + 1 + 16, or 18 times as heavy as the hydrogen atom. So an amount of water holding the above number of water molecules weighs 18 grams. This special number of atoms or molecules is called the *Avogadro number*, named after the eighteenth-century Italian physicist Emilio Avogadro whose careful studies of chemical reactions between gases led eventually to a determination of the number. A weight of any substance holding an Avogadro number of molecules (or atoms in the case the substance is an element) is given a special name by chemists. They refer to it as one *gram mole*, or simply one *mole*, of the substance. One mole of water is 18 grams of the substance. One mole of carbon dioxide weighs 44 grams. We therefore see that the carbon dioxide molecule is 44 times as heavy as the hydrogen atom, because an Avogadro number of hydrogen atoms weighs one gram while the same number of carbon dioxide molecules weighs 44 grams.

The mole is the chemist's unit of quantity measure. It differs from ordinary weight units like pounds or grams in that it takes differences in weight between different kinds of atoms and molecules into consideration. A mole of any substance always holds an Avogadro number of basic particles—atoms or molecules. But one mole of one substance does not weigh the same as one mole of another substance. That is because different atoms and molecules do not weigh the same. Also, n moles of a substance contains n times the Avogadro number of basic particles. So 36 grams of water, or two moles of the substance, has twice the Avogadro number of molecules.

Chemists measure weights in moles when analyzing the different kinds of DNA. If the four-nucleotide theory in its modified form was correct, the breakdown of one mole of any kind of DNA into its component four-nucleotide substances should give equal numbers of moles of each of these substances. Actually, the weights of DNA biochemists dealt with was much less than one mole of the substance. Yet any number of moles of DNA should yield equal numbers of moles of the four component substances according to the theory.

But such was shown *not* to be the case by experiment. The numbers of moles of each of them from DNAs in many diverse organisms were not the same by far; the modified four-nucleotide theory had to be wrong. It was becoming apparent that the general DNA polynucloetide chain could have any number of each of the four base units arranged in any order along its length, while complex calculations showed that if such was the case DNA was more than able to carry instructions for the making of all the varied peptide chains found in living matter.

But there are only four bases along the DNA chain and up to 20 amino acid units along the peptide chain. Many more proteins than DNAs could arise. There still did not at first seem to be enough variety in the DNA blueprint. But the number of protein chains in the average cell in a triffling fraction of all those theoretically possible on basis of protein structure. The DNA chains in an organism are more than able to make the actual proteins needed by the cell. Their number is nowhere near as large as the total number that can be made.

So both through experiment and theory, biochemists came to see that DNA not only could, but did, carry the hereditary units envisioned by Mendel and others after him, although this alone told them nothing about the manner in which the DNA polynucleotide chain held these units. Also, they remained in the dark about the nature of the chemical blueprint the DNA chain contained. Obviously there were DNA chains of many different lengths in the chromatin of cell nuclei; that much was clear. In some way a chemical message was sent from these chains to the ribosomes of the cytoplasm, the sites of protein building, which had to entail certain chemical directives (or orders) for the making of each peptide chain needed by the cell. It was becoming apparent that RNA was somehow involved in carrying out these directives. The details of the process were still unclear at the time of these speculations, the early 1950s. Ribosomes contained large amounts of RNA; that much became certain.

In fact, a crucial step toward getting a clearer picture of cell chemistry was taken in the early 1950s, after certain patterns had been noted in the compositions of DNAs from many different sources. One of these was that the total number of adenine units was the same as the total number of thymine units in any DNA molecule. Also the total number of guanine units equaled the total number of cytosine units. Now adenine and guanine are purines, while thymine and cytosine are pyrimidines. The total number of purine units in any DNA molecule must therefore equal the total number of pyrimidine units. What could enforce such restrictions on the base units of any DNA molecule? No explanation of this phenomenon presented itself as long as the DNA molecule was seen as a single polynucleotide chain, because, in that case, no mechanism stood out to account for these odd equalities observed among the numbers of different base groups in the molecule. At least no biochemist could come up with any productive ideas about the matter on that model.

The reasons for such composition patterns came in 1953 through DNA research of a different kind. An English physicist M. H. F. Wilkins was probing DNA structure by means of X-rays. The physics of X-rays, their production, and how they are used to study molecular structure would take things far beyond the scope of this account. Wilkin's X-ray investigations of DNA crystals showed that NDA had a double-stranded molecule—a molecule consisting of two polynucleotide chains wrapped around each other. Nearly all RNA molecules, on the other hand, were found to be single stranded. Also no such composition patterns had been found among the base units of RNA. A small minority of RNA molecules are double-stranded—not very many, however.

THE DOUBLE HELIX

Wilkin's findings provided the first clues to an explanation of the strange composition patterns among DNAs. Further X-ray studies showed that, first of all, the two chains of the DNA molecule were not straight, with one lying beside the other. Instead each had the shape of a long, coiled spring, a figure called a *helix*. The axis of a helix is a straight line passing through its middle along its length about which the helix winds (FIG. 3-3). The DNA molecule came to be seen as two, helical, polynucleotide chains winding around the same axis and joined to one another at the center of the molecule by means of a weak attraction between the base groups. The sugar and phosphate groups of each chain are located in the outer portions of the long molecule. The base groups on one of the chains are

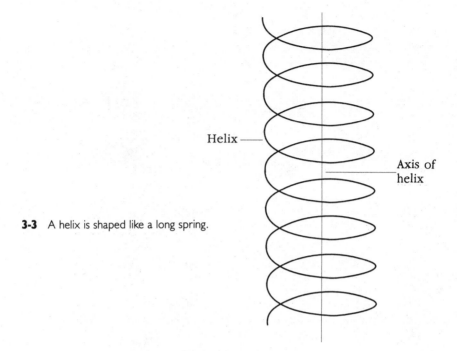

Helix ——

Axis of
helix

3-3   A helix is shaped like a long spring.

bound to those of the other at the common axis of the two chains at the center of the molecule. Figure 3-4 illustrates this. X-ray data also implied that the sugar-phosphate backbones of the helical chains are a fixed distance apart in all DNA molecules.

Each base group at the center of the double helix on one of its chains is bound to one on the other chain. For any two bound base groups at the core of the molecule, there are three different possibilities:

1. a purine group may be bound to another purine group
2. a pyrimidine group may be bound to another pyrimidine group, or
3. a purine group may be bound to a pyrimidine group

Whichever of these three alternatives is correct, it must be consistent with one prime experimental fact: The distance between the sugar-phosphate backbones of the two polynucleotide chains must be the same everywhere along the molecule; it cannot vary anywhere. This had to be the case according to X-ray data on DNA structure. How does each of the three base pairing alternatives stand up to this requirement? Both purine molecules, adenine and guanine, consist, for the most part, of two rings of atoms of the same size. So if a purine base group is bound to another purine base group all along the DNA molecule, there would be four rings of atoms all of the same size between the sugar-phosphate backbones at all points of the molecule. That would make the distance between the

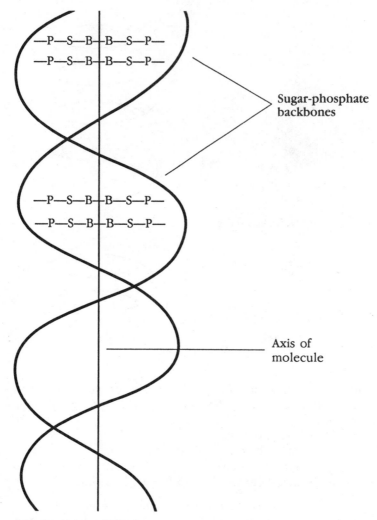

—P—S—B—B—S—P—
—P—S—B—B—S—P—

Sugar-phosphate
backbones

—P—S—B—B—S—P—
—P—S—B—B—S—P—

Axis of
molecule

3-4 The DNA double helix.

backbones everywhere the same. On the other hand, the two pyrimidine molecules, thymine and cytosine, are mainly composed of a single ring of atoms of the same size in both. Therefore, the fixed-distance requirement is met if a pyrimidine group is bound to another pyrimidine group everywhere along the molecule. Both alternatives agree with the fixed-distance requirement. Yet both do not stand up to another basic experimental fact of DNA composition: all forms of DNA contain both purines and pyrimidines, and not just purines or just pyrimidines. This was shown clearly by the composition patterns mentioned earlier.

This leaves only the third possibility. There are three rings of atoms between the sugar-phosphate backbones. The atomic rings of both purine molecules are the same size. The single atomic rings in the two pyrimidine molecules are of the same size also, making the three-ring combinations between the backbones all of the same length. Both the fixed-distance and the composition requirements are met.

The question of which purine pairs up with which pyrimidine remains open. This problem was not hard to figure out, though, since it was known by experiment that the total amount (in moles) of adenine (a purine) is the same as that of thymine (a pyrimidine) in all DNAs. Thus the total number of adenine units in the molecule had to be the same as the total number of thymine units, while the same holds for the guanine and cytosine units. These observations, along with the fact that all DNAs contain both purines and pyrimidines, could only mean that adenine is always bound to thymine. Also guanine is always bound to cytosine. These combinations are symbolized by the short notations A−T and C−G. A is said to be the *complement base* of T, and T is the complement base of A. The same relationship holds between the bases C and G also. These pairing rules have certain implications. One is that if we know the order of the bases on one of the DNA chains, we can easily figure out the base order on the other chain. The base groups on one of the polynucleotide chains can come in any order. But once we know that order, the order of the bases on the other chain is established by the A−T and C−G pairing rules.

The nature of the bonding between the base groups of a base pair is not the same as that of ordinary chemical bonds between atoms in molecules. It is somewhat weaker. But it is able to hold the two chains together against the thermal motion of the molecule.

Imagine a small, simple DNA molecule of eight nucleotides per chain. Call one of its short, interwinding chains J, and its other chain K. With the base pairing rules A−T and C−G we can easily get chain K by noting that the first base group on chain J, an A group, must be bound to a T group on chain K, making the first base group on that chain a T group. In the same way, the second base group on J, a G group, must be bound to a C group on chain K. This simple DNA molecule can be depicted as in FIG. 3-5, where the helical shapes of the short chains of the molecule are not shown.

3-5 Complementary chains in DNA. In this example, chains J and K are complements of each other, and they join as shown.

Chain J	Chain K	
A	T	A—T
G	C	G—C
C	G	C—G
T	A	T—A
G	C	G—C
C	G	C—G
A	T	A—T
T	A	T—A

Next consider a much longer DNA molecule of thousands of nucleotides per chain. If the base arrangement on one chain of the long molecule starts out as ATGCTTCAG . . . , the A−T and C−G pairing rules tell us that the other chain begins in the fashion TACGAAGTC . . . .

The theory picturing the DNA molecule as made up of two helical, interwinding, polynucleotide chains with the bases following the A−T and C−G pairing rules is known as the Watson-Crick model. The model is named after its inventors F. H. C. Crick, an Englishman, and J. D. Watson, an American. It is the best theory of DNA structure at present. No flaws have been found in it since it was proposed in 1953.

After the structure of these important biochemicals were derived, biochemists turned their attention to another intriguing problem: that of chromosome replication. Because DNA had been shown to be the genetic material in chromosomes, and because replication was one of the chief features of chromosomes, it seemed fair to conclude that the DNA molecule could somehow reproduce itself. But how? It is to this question we now turn.

Chapter **4**

The chromosome in genetics

Much has been said about DNA being the genetic material. But such knowledge did not directly help life scientists in their quest to understand precisely what the gene is. Genes must exist on chromosomes; that much had been learned. However, it was quite clear that the chromosome itself could not be the gene.

THE GENE AND THE CHROMOSOME

Each human body cell hold 46 chromosomes, while the average human being has thousands of traits, many of which are known to be controlled by a single gene, since they are passed to offspring in the same way as the traits Mendel studied in pea plants. There was no escaping the conclusion that each of the 46 chromosomes held many genes. Besides, the number of traits other organisms—even one-cell types—had when compared to their fixed number of chromosomes gave the same results. It was not exceptional for a single chromosome to hold hundreds of genes. Chromosomes are made of chains of DNA. Therefore, each gene had to be a small part of the polynucleotide chain of some DNA molecule in some chromosome. What was the exact nature of these parts, and how many base groups did they hold on the average? As geneticists and biochemists set out to answer these questions, they addressed the central problem of molecular biology: that of how the base arrangements along the DNA chain determine the protein chains needed by the cell.

Toward that end, geneticists in the first third of the twentieth century began to map the chromosomes found in small fruit flies and one-cell organisms. Such mapping showed where specific genes like those for eye color or wing shape of the fly were situated on the chromosomes in the nuclei of its body cells. Genetic mapping has given much valuable information about the nature of the once mysterious genes proposed by Mendel. Let's look at some of these interesting and fruitful developments.

A CLOSER LOOK AT CHROMOSOMES

That DNA is the genetic material is a relatively recent discovery. It goes back to 1944, the year Avery and his colleagues did their work with E coli bacterial cells. But chromosomes had been known long before that. The presence of these long bodies in cell nuclei was clearly established in the middle of the nineteenth century, when the dye industry in Germany, and better microscopes, had come into being. Mendel completed his work with garden peas around 1865. But, at that time, not much was known about cell organelles and the part each type of organelle plays in cell processes. Good microscopes, and dyeing chemicals that made these structures visible, had been lacking. Chromosomes had been seen, but not much was known about them. So a lot of speculation, backed by little experimental observation, was put forth regarding their purpose in the cell. Some investigators thought that chromosomes behaved like Mendel's genes. To understand how they came to that conclusion, let's look at Mendel's theory more closely.

Mendel looked at the passing of one single trait—for example, seed shape, seed color, or plant height—from parent plants to offspring. From these experiments he arrived at his gene theory. But in what manner were genes passed to offspring? Were those for a given trait passed along independent of those for other traits, or did different genes affect one another on being passed to the seeds of offspring? To answer these questions, Mendel devised a series of simple, but laborious, experiments. He crossed pure plants differing in two or more traits. He then examined their offspring for the traits and the way the traits were distributed.

In one experiment he crossed pure plants, having tall vines and round seeds, with pure dwarfed ones having wrinkled seeds. From his previous work he knew that tallness and round seed shape were dominant traits, while dwarfism and wrinkled seed shape were recessive traits, so that it came as no surprise that all plants coming out of this cross were tall with round seeds. But, according to his gene theory, this first generation were hybrids since they came from parents differing in both traits. Their gene pairs for height and seed shape each held two kinds of genes. One was dominant and the other recessive. The height gene pairs in their cells had the genetic composition Td. The seed shape pairs had the composition Rr. (R is the dominant gene for round seed shape and r is the recessive one for wrinkled seed shape.) So tallness and round seed shape accompanied each other in all first generation plants, as the theory implied they should. Would they do so in the second generation of plants? What kind of plants would come out of a cross between the first generation plants in this experiment? Mendel self-crossed them to find out.

The theory stated that genes for a given trait, like height or seed shape, should be passed to seeds independent of each other. The first generation plants in this example had two genes in their cells for the trait of height: a T gene and a d gene. On being crossed, each plant among them was as likely to give a T gene as a d gene to a given seed. At the same time, each plant was as likely to give an R gene to a seed as an r gene to the same seed. Mendel's assumption that the height genes and seed shape genes, for instance, were passed to seeds independently also seemed to imply that if a parent plant gave a T gene to a seed, such did not affect the plant's ability to give either an R gene or an r gene to the same seed with equal likelihood. If it gave a d gene to a given seed, that should likewise not affect its ability to give either of the seed shape genes (R or r) with equal probabilities to the same seed.

At this point, it is instructive to go back to the case in which Mendel crossed two pure plants differing only in height. In that case, the first generation were the tall, hybrid plants having the genetic composition Td in their height gene pairs. When these tall hybrids were self-crossed, each parent plant had an equal likelihood of giving a T gene or a d gene to a given seed. Call one parent of each pair of crossed plants A and the other B. In FIG. 4-1, the horizontal row at the top consists of the possible height genes A can give to a seed—a T gene or a d gene—while the vertical row

4-1   The crossing of genes. The parents are A and B. Tallness is represented by T; dwarfism by d.

to the left consists of the height genes B can give to the same seed—also a T or a d gene. The symbols in the squares are possible genetic compositions of the height gene pairs of seeds coming out of the cross. For example, the square at the upper left represents the case where both A and B give a T gene to the seed. The square at the upper right represents the case where A gives a d gene and B a T gene. In the first case, the seed would have the genetic composition TT in its height gene pair, and would therefore give a pure tall plant on sprouting. In the second case, the seed would have the genetic composition Td. In that case the seed would give a tall, hybrid plant on sprouting. All the plant's body cells would have the height gene pairs of the composition Td. The square at the lower left represents the case where A gives a T gene and B a d gene to a given seed. The height gene pair of the seed will again have the genetic composition Td, and will again give a tall, hybrid plant. The square at the lower right depicts the case in which both A and B give a d gene to a seed. Then the seed has the genetic composition dd. It therefore gives a pure dwarfed plant. There is no such thing as a hybrid, dwarfed plant, since d is a recessive gene.

Because each parent plant is as likely to give a T gene as a d gene to a given seed, the four cases represented by each box in the above diagram occur with the same frequency, on the average, among a large number of seeds. In other words, a seed of the composition TT should arise, on the average, one quarter of the time in such a cross. One with the composition dd should arise that often also. So, out of a large number of seeds coming out of the cross under consideration, around one quarter will have the height gene pair TT and about one quarter the pair dd. But the height gene pair Td occurs in two of the four boxes, or cases. It will thus occur in seeds about half the time. Around half of the seeds in a large number coming out of the cross will have the height gene pair Td. Both seeds having the height gene pair TT and those with the pair Td give tall plants. Therefore, on basis of our assumption that each parent hybrid has a $50/50$ chance of giving a T gene or a d gene to any seed, we expect that about three quarters of the second generation plants will be tall and about one quarter dwarfed. That, recall, is just what Mendel found.

A similar diagram for the inheritance of seed shape is shown in FIG. 4-2. A cross between round, seeded, first generation hybrids should give a second generation in which about three quarters of the plants are round seeded and about one quarter wrinkled seeded. That, too, is what Mendel found.

A

	R	r
R	R R	R r
r	R r	r r

B

4-2 Inheritance of seed shape. Parents are A and B. Roundness is represented by R; wrinkliness by r.

The ratio of about three offspring having the dominant trait to one having the recessive trait follows from the fact that each parent plant has an equal chance of giving either gene to a given seed.

These patterns aid in anticipating what Mendel believed should happen when he crossed the tall, round seeded hybrids with each other. But, at this point, he made another assumption: genes for plant height are passed to seeds independent of those for seed shape. There should be no tendency for tallness and round seed shape to accompany one another into the second generation.

Consider the height gene and seed shape gene pairs taken together. Let the genetic composition of both pairs taken together be designated by the symbol HS, where H is the genetic composition of the height gene pair of the plant and S is the same for the seed shape pair. The original pure plants that were tall with round seeds had the genetic composition TTRR with regard to the two traits being considered, because they held two genes for tallness in their height gene pair and two for round seed shape in their seed shape pair. Likewise, the original pure plants that were dwarfed with wrinkled seeds had the composition ddrr for the two traits.

In the first case H = TT and S = RR. In the second case H = dd and S = rr. The first generation hybrids had the composition TdRr in the two traits. In their case H = Td and S = Rr. Mendel's second assumption meant that the two genes in the H pair of a seed were not in any way determined by the genes in the S pair in the same seed. Similarly, the genes in the S pair would not depend on what genes were in the H pair in the seed. Thus, when two of the hybrids were self-crossed, each parent could contribute a T gene to the H pair of a given seed coming out of the cross, or one of them a T gene and the other a d gene, or both could contribute a d gene, while the same comments apply to the S pair. So each hybrid parent could make the genetic contributions [T|R], [T|r], [d|R], or [d|r] to the combined gene pairs HS in any of their seeds (FIG. 4-3). Parent A can make four kinds of genetic contributions HS as can B. This makes 16 different combined pairs (or 16 different HS's) altogether, given by the symbols in the boxes. In the cases

A

		T R	T r	d R	d r
	T R	TT RR	TT Rr	Td RR	Td Rr
	T r	TT Rr	TT rr	Td Rr	Td rr
B	d R	Td RR	Td Rr	dd RR	dd Rr
	d r	Td Rr	Td rr	dd Rr	dd rr

4-3 Offspring of hybrid parents. Symbols are discussed in text.

$$\overset{1}{\text{HS}} = \overset{}{\text{TTRR,}} \quad \overset{2}{\text{TTRr,}} \quad \overset{2}{\text{TdRR,}} \quad \overset{4}{\text{TdRr}}$$

the offspring are tall with round seeds, because of the dominance of the T and R genes. The small numbers above each given HS combination indicate the number of times that particular composition of HS occurs in the 16 cases, or is the number of boxes holding that particular composition of HS. In the present case, nine out of the 16 HS compositions give tall plants with round seeds.

When

$$\text{HS} = \overset{1}{\text{TTrr,}} \quad \overset{2}{\text{Tdrr}}$$

the offspring are tall with wrinkled seeds. There are three such cases in the 16.

In another instance, when

$$\text{HS} = \overset{1}{\text{ddRR,}} \quad \overset{2}{\text{ddRr}}$$

the offspring are dwarfed with round seeds. There are again three such cases in the 16.

In one case of the 16

$$\text{HS} = \overset{1}{\text{ddrr}}$$

and the offspring are dwarfed with wrinkled seeds.

Therefore, if the height and seed shape genes are passed to seeds independent of one another, as Mendel supposed, the following proportions of the four kinds of offspring should roughly be found in the second generation:

Tall plants with round seeds: $9/16$ of the total
Tall plants with wrinkled seeds: $3/16$ of the total
Dwarfed plants with round seeds: $3/16$ of the total
Dwarfed plants with wrinkled seeds: $1/16$ of the total

if, of course, one considers a very large number of offspring. These were approximately the proportions of the four kinds of offspring plants in Mendel's second generations. That tended to show, in his opinion, that genes for height are passed to seeds independent of those for seed shape. The traits of the two genes showed no particular tendency to stay together in the offspring from generation to generation.

What if the two kinds of gene always stayed together in being passed to seeds? Then Mendel would have found only tall plants with round seeds and dwarfed ones with wrinkled seeds in all generations. No cases where dwarfism accompanied round seed shape, or where tallness accompanied wrinkled seed shape, would be found in any generation.

From his results, Mendel concluded that different kinds of genes were passed to offspring independent of each other, and in a totally random way.

Mendel had discovered two things about the behavior of genes: those for a given trait come in two kinds, one dominant and the other recessive, while those for different traits were passed to offspring independent of one another. It was easy to determine whether two genes were passed independently, by examining the proportions of the four different kinds of offspring in the second generations in a breeding experiment. If the proportions were

$$9/16, \ 3/16, \ 3/16, \ 1/16$$

the genes had to be passed to offspring independent of each other. Mendel was satisfied that he found just these proportions in all his experiments with the seven traits of the pea plant he had studied. Each trait was passed to offspring independent of the other six. No two showed any tendency to stay together from generation to generation. He called this behavior the law of independent assortment of traits. Different traits assorted independently and showed no preference to stay with others in offspring.

Was there a cellular explanation for these findings? In Mendel's time there was much theory but no valid observations to account for what he had uncovered. This, of course, began to change when better microscopes and dyes to color cellular organelles came into being.

SOMATIC AND GERM CELLS

Twenty years after Mendel's work, the zoologist August Weismann came up with the idea that there are two kinds of cells in many organisms. One type are the ordinary body cells, today called somatic cells, making up various tissues and organs. The other type, though much fewer in number in the organism, but very important, he named germ cells. Germ cells, he maintained, are continually being produced in the reproductive organs—the ovaries of the female and the testes of the male—in organisms that have sex, and today these cells go by the names of egg and sperm cells. On mating, one egg cell from the female unites with a sperm cell from the male to form the first cell of the newborn individual. According to Weismann, all genetic information for the making of the new individual was contained in the germ cells. Other body cells played no part in reproduction whatever, except for a small number of them that form the germ cells.

At the time of Weismann's speculations, dyes, that colored cell organelles so that they could be seen under the microscope, had been discovered and put to work. Body cells and germ cells could be examined more clearly under the microscope. That is how chromosomes were studied in more detail. When certain coloring agents were applied to the cell, a material was found in the nucleus that became dark red, or crimson, in their presence. This material became known as chromatin. As previously mentioned, this material shapes itself into long, rod-shaped structures, the chromosomes, before cell division. From 1870 to 1920 biologists watched cell division of all kinds under their microscopes. One peculiar occurrence caught their attention: chromosomes were the only cell organelles that showed an orderly and visible motion during cell division, while other organelles did not, at least in the same fashion, with half the total number of chromosomes going to one of the daughter cells and half to the other. Chromosomes were clearly particle-like bodies that were passed from parent to daughter cells in a systematic way. A given chromosome could just as well go to one daughter cell as to the other, and had an equal chance of ending up in either. The microscope revealed that germ cells had half as many chromosomes as normal body cells. When male and female conceive, the first cell of the new individual formed by the union of the egg and sperm cell was endowed with the normal number of chromosomes, half of which came from the female's egg cell and half from the male's sperm cell.

The fact that chromosomes are particulate bodies that are passed in a random way to daughter cells in cell division led some scientists to think that they could be the genes Mendel had speculated about. Genes, too, had been envisioned as particles of some kind that are passed to offspring in a completely random fashion. Chromosomes were identified as Mendel's genes by some observers. This seemed like an accurate idea at first, since, in the last half of the nineteenth century, precise chromosome counts in organisms had not been made. Each body cell was thought, in this picture, to hold two chromosomes for each body trait. One chromosome of each pair would be for one aspect of a trait, while the second chromosome would be for the other aspect. But each germ cell, having half the normal amount of chromatin in their nuclei, would hold one or the other of each kind of chromosome. Germ cells are made in large numbers in an organism's reproductive organs. So, by Mendel's law of independent assortment, half the total number of germ cells should have a chromosome for one aspect of a given trait and half for the other aspect, if the chromosome is identical to the gene. In pea plants, for example, each body cell would have two chromosomes for plant height in its nucleus. Both of them could be for tallness, so that the plant would be pure tall. Or one of them could be for tallness and the other for dwarfism, making the plant a tall hybrid plant, because of the dominance that the chromosome for tallness would have. Finally, both height chromosomes could be for dwarfism, giving a purely dwarfed plant.

Thus, if the gene is seen as the chromosome, each body cell of the pea plant would hold a pair of chromosomes for each of the traits studied by Mendel, both of which could be for one aspect of the trait, or one for one aspect and one for the other, although when each germ cell is formed from special body cells, it would get one of the chromosomes of each pair for one trait aspect from one of these special body cells.

Consider the height chromosomes in this theory. In a pure dwarfed plant, all body cells would hold two chromosomes for dwarfism, while the germ cells would hold only chromosome for dwarfism. A pure tall plant would have body cells holding two chromosomes for tallness, while its germ cells would hold one tallness chromosome. When a germ cell from a pure tall plant united with one from a pure dwarfed plant to give the first cell in the seed of a new plant, that cell would hold one tallness chromosome and one dwarfism chromosome, as would all cells arising from that seed in the new plant. All plants of this first generation would have one chromosome for tallness and one for dwarfism in their body cells. They would all be tall because of the dominance of the tallness chromosome.

Germ cells from these first generation hybrids would be of two kinds. Half of them would get a tallness chromosome and half a dwarfism chromosome. So when two first generation plants are crossed, their germ cells unite at random in four possible ways to form the first cells of off-spring plants.

Any two germ cells that unite to form a seed could each contain a tallness chromosome. That is one of the four possibilities. Another is that the germ cell from parent 1 could hold a tallness chromosome and that from parent 2 a dwarfism chromosome. A third possibility is that the germ cell from parent 1 holds a dwarfism chromosome and that from parent 2 a tallness chromosome. The fourth possibility is that the germ cells from both parents hold a dwarfism chromosome. On basis of this chromosome theory of the gene, all four cases should happen with equal frequency, making about one fourth of the seeds coming out of the cross with two tallness chromosomes, so that about one fourth of the offspring plants will be pure tall. The second and third possibilities would be represented by about half the seeds. So about half the total number of offspring would be tall, hybrid plants. About one quarter of the seeds would hold two dwarfism chromosomes, making about one quarter of the second generation pure dwarfed. So, out of a large number of second generation plants, about three quarters should be tall and about one quarter dwarfed.

PASSING OF TRAITS

The assumption that the gene is the same as the chromosome explained Mendel's first experiments quite well. But what did it have to say about his other experiments, in which he followed the passing of two or more different traits into the second generation of plants? Let's look at this problem to see why the idea of the chromosome being the gene might have been attractive to some early investigators.

In the original pure plants that were tall with round seeds, each body cell would hold two chromosomes for tallness and two for round seed shape. As these plants were crossed to give a first generation, their germ cells would each hold a chromosome for tallness and one for round seed shape. On the other hand, the original dwarfed plants with wrinkled seeds would have germ cells holding a chromosome for dwarfism and one for wrinkled seed shape. The germ cells of the pure tall plants with round seeds could be given by the symbol T|R. Those of the pure dwarfed ones with wrinkled seeds could likewise be represented by the symbol d|r. Therefore, when the two pure types crossed to give a first generation, their germ cells united at random to give seeds of the genetic compositions Td|Rr. This produced tall hybrid plants with round seeds, because of the dominance of the T and R genes or chromosomes, even though the body cells of these plants would hold a chromosome for dwarfism and one for wrinkled seed shape.

The first generation hybrids have germ cells of the four genetic compositions: T|R, T|r, d|R, and d|r. These germs cells unite at random when the first generation plants are crossed to give a second generation. There are 16 possible combinations of the germ cells when they unite to give seeds of the second generation. Each first generation hybrid plant makes germ cells of the above types in equal numbers. That much could be assumed. One fourth of the total number of germ cells each parent plant produces have the genetic composition T|R, while another quarter of them have the composition T|r, and so on. When the first generation plants are crossed, their germ cells unite in the following way:

T|R	+	T|R →	TT|RR	gives tall round seeded plants
T|R	+	T|r →	TT|Rr	gives tall round seeded plants
T|R	+	d|d →	Td|RR	gives tall round seeded plants
T|R	+	d|r →	Td|Rr	gives tall round seeded plants
T|r	+	T|R →	TT|Rr	gives tall round seeded plants
T|r	+	T|r →	TT|rr	gives tall wrinkled seeded plants
T|r	+	d|R →	Td|Rr	gives tall round seeded plants
T|r	+	d|r →	Td|rr	gives tall wrinkled seeded plants
d|R	+	T|R →	Td|RR	gives tall round seeded plants
d|R	+	T|r →	Td|Rr	gives tall round seeded plants
d|R	+	d|R →	dd|RR	gives dwarfed round seeded plants
d|R	+	d|r →	dd|Rr	gives dwarfed round seeded plants
d|r	+	T|R →	Td|Rr	gives tall round seeded plants
d|r	+	T|r →	Td|rr	gives tall wrinkled seeded plants
d|r	+	d|R →	dd|Rr	gives dwarfed round seeded plants
d|r	+	d|r →	dd|rr	gives dwarfed wrinkled seeded plants

The second generation sprout from 16 different kinds of seeds. The seeds yield four types of plants. Nine of the seed types give tall, round seeded plants, while three give tall, wrinkled seeded ones. Three give dwarfed, round seeded plants, while one gives dwarfed, wrinkled seeded ones. But these are just proportions of the four types of plants Mendel found in his experiments.

These patterns of pea plant inheritance are based on three assumptions:

1. Genes—or chromosomes in this theory—for a given trait are passed unblemished from generation to generation.

2. Genes, or chromosomes, are given to germ cells independent of other genes in a totally random way.

3. The gene is the same as the chromosome seen under the microscope.

These assumptions explain Mendel's main findings quite well. But were they generally valid?

To this day, no one has found fault with assumption 1 above. The idea that a particle-like body, the gene, is the carrier of each trait an organism displays has stood the test of time and new findings have not cast the least doubt on the matter. But the third assumption, that the gene and the chromosome are the same, was only a *hypothesis*—an intelligent guess or hunch the scientist feels might explain his findings—in the early twentieth century. But to some it seemed like a sound one. It had some very obvious drawbacks, however. One of these was clear to anyone who pondered the problem. Consider a small fruit fly. One such fly was known to have eight chromosomes in its body cells. If the gene and the chromosome were the same, it would be reasonable to conclude that the little fruit fly should have no more than eight traits which, of course, was preposterous. At least several hundred traits were easily observed in the fly. So each of the eight chromosomes had to regulate many traits, and hence, each must hold many specific genes.

Humans are an even better example. An average human being is known to possess at least thousands of traits, many of which are controlled by a certain gene. So there must be at least hundreds of genes in the chromatin of human body cells. Yet studies revealed only 46 chromosomes in the nuclei of human body cells. Thus each chromosome had to have at least hundreds of genes.

By the turn of the century it was clear that the gene and the chromosome could not be identical. All organisms studied—even simple, one-celled ones with one or two chromosomes—showed numbers of discernible traits that were far in excess of the number of chromosomes they held. The idea that the gene and the chromosome were identical was a simple and attractive idea. A simple cytological explanation of heredity would then be available; but the idea was doomed to failure.

But how could the idea explain Mendel's work if it was false? It all had to do with the second assumption that different genes are passed from body cells to germ cells independent of each other, a theory that turned out to be true only in certain cases. Mendel's work with pea plants was one of those cases. Traits like tallness and round seed shape are passed from generation to generation independent of each other because their genes happen to be on different chromosomes. There are seven chromosomes of various lengths in pea-plant cells. The gene for height is on one of them. That for seed shape is on another. However, the chromosome holding the height gene holds many others besides, as does that holding the seed shape gene. Since the two genes are on different chromosomes, they, like their chromosomes, are passed to germ cells and thus to seeds, independent of each other, leading to the proportions

$$9/16, \quad 3/16, \quad 3/16, \quad 1/16$$

of the four different types of plant in the second generation.

But what should have happened if the two genes were on the same chromosome in pea plant cells? Then, the original tall plants with round seeds, as well as the original dwarfed ones with wrinkled seeds, would have germ cells of the genetic compositions [T|R] and [d|r], although now the genes T and R (and also d and r) are on the same chromosome in these cells. When the two kinds of plants are self-crossed, their germ cells unite in the fashion

$$[T|R] + [d|r] \rightarrow [Td|Rr]$$

to give a first generation of tall plants with round seeds that are hybrids. But, then, what kind of germ cells would these first generation plants have? If the height gene and the seed shape gene are on different chromosomes, the plants would have four different types of germ cells with the compositions [T|R] , [T|r] , [d|R] , and [d|r] because there would be no tendency for the R gene (or the r gene) to follow the T gene into a given germ cell. A germ cell that would get a chromosome with a T gene would have equal chances of getting another that holds either an R gene or an r gene. It is that simple. When chromosomes are passed at random to germ cells, each of the four different genetic compositions above would have equal likelihoods of turning up in a given germ cell. On the average, each composition will turn up in a germ cell one quarter of the time. But, if the genes T and R are on the same chromosome, they follow each other into a germ cell in the first generation, as will d and r. Then, the first generation plants would have germ cells of only two compositions, not four: [T|R] and [d|r].

When these plants are self-crossed to give a second generation, the above germ cells made in equal numbers in each parent plant, would unite according to the scheme shown in FIG. 4-4, to give a second generation that are either tall with round seeds or dwarfed with wrinkled seeds in the proportions of about three of the first type to one of the second. No tall plants with wrinkled seeds or dwarfed ones with round seeds would occur in that generation. Tallness and round seed shape would stay together in all generations, as would dwarfism and wrinkled seed shape.

LINKAGE AND MIXING

Traits that stay together generation after generation, because their genes are on the same chromosome, are said to be *linked*. A set of genes on a

Germc cell of
one parent

Germ cell of
another parent

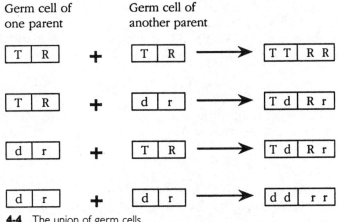

4-4   The union of germ cells.

given chromosome in an organism is therefore called a *linkage group*. In reality, traits like height and seed shape in pea plants belong to different linkage groups, so that they do mix in the second and later generations in contrast to their behavior in our make-believe example. Mendel's work, in fact, showed that such plants have at least seven linkage groups, the traits of which are passed to offspring independent of each other, while microscopic investigation showed that pea plants have seven different kinds of chromosomes, each of which holds the genes for one of the seven linkage groups.

So the number of linkage groups of genes possessed by an organism is the same as the number of different chromosomes found just before cell division in its body cells. This is because each group has genes on a certain chromosome. In pea plants there are seven sets of linked traits, or seven linkage groups. There must be a chromosome for each group, making seven kinds of chromosomes of different lengths and genetic compositions. Humans have 23 different chromosomes. Therefore, they have 23 sets of linked traits that tend to be passed to their children together. The fruit fly *Drosophila* has four different types of chromosomes. It thus has four linkage groups of traits. The study of them has shed much light on our current knowledge of chromosome behavior.

Our make-believe example with the pea plant is a fictitious example in another way. For one thing, there is no mixing of the two linked traits in the second generation. That seldom occurs in real life studies of linked traits. Often there is some mixing of linked traits, but not to the extent Mendel's laws of independent assortment require.

Either two traits are linked or they are not. If they are linked, with genes on the same chromosome, they should stay together in all individuals in all generations with no mixing. Also, their proportions in any generation should be those dictated by Mendel's law of independent assortment applied to chromosomes. And if the traits are not linked, they should mix in the second generation of a breeding experiment in a totally random way, while the proportions of the four different types of off-

spring should be those indicated by Mendel's law of independent assortment for individual genes. What is often found in reality is somewhat
between these extremes. An example would be if Mendel found the proportions

$$11/16, \quad 1/16, \quad 1/16, \quad 3/16$$

all of the four types of offspring with regard to height and seed shape in
his second generations instead of the proportions

$$9/16, \quad 3/16, \quad 3/16, \quad 1/16.$$

What is the explanation for such odd cases?

As we have said, body cells and germ cells differ in one important
way: body cells have twice the number of chromosomes (or twice as
much chromatin) as germ cells and have two chromosomes of each kind,
while germ cells have only one of each kind. Cytologists, or biologists
who study cells and their inner structure, have thoroughly observed cell
division and germ cell formation under the microscope. They have
learned much about both processes. In ordinary cell division the chromosomes just duplicate. An equal number are passed to each daughter cell,
with the new chromosomes being exact copies of the original chromosomes in the mother cell. No new kinds of chromosomes, as far as genetic
makeup is concerned, are normally created in the process. So the chromosomes of the daughter cells are identical to those of the mother cell.
But something a little different takes place in germ cell creation from
body cells. Let's look at the process in some detail; it is both interesting
and important in explaining the above-mentioned deviations from Mendel's laws.

CROSSOVER

Certain body cells in the sex organs divide not to give more cells like
themselves, but germ cells, holding half the normal number of chromosomes. These special body cells do not divide only once, but twice, in
doing so. Consider one of these special body cells. It, like any other body
cell, holds the normal number of chromosomes, which come in pairs.
Each chromosome pair holds two chromosomes of the same length—
except for two special ones called sex chromosomes—which are the same
gene to gene, while one chromosome of the pair is inherited from one of
the parents of the organism and the other from the other parent. Just
before this cell divides, each chromosome makes a duplicate of itself. Just
prior to chromosome duplication, the chromosomes of each pair attach
themselves to each other while each chromosome takes advantage of cell
chemicals to make another chromosome like itself, making a total of four
chromosomes bound to each other where only two existed before. But at
this stage of the process more than mere duplication often takes place.
Two of the chromosomes in the foursome often exchange parts to some

extent to give two new chromosomes of the same length. The two new chromosomes, however, have different genetic makeup from each other and from the original two chromosomes. The four chromosomes then separate. This phenomenon (FIG. 4-5) is known as *crossover*. Crossover refers to the fact that two chromosomes of the foursome exchange parts or genes. It does not occur in normal cell division. In crossover, blocks of genes are exchanged between the two chromosomes. After this entire process, the cell is left with twice the normal number of chromosomes until it divides. When the cell divides half the chromosomes go to one daughter cell and half to the other. This happens in a perfectly random way. Thus the two daughter cells that result have the normal number of chromosomes with one crucial difference: they also have chromosomes of a different genetic composition than those of the body cells from which they came because of crossover, or gene exchange. The two daughter cells then divide without chromosome duplication. This results in four germ cells with half the normal number of chromosomes. Such creation of germ cells from special body cells is known as *meiosis*, while normal cell division without crossover among chromosomes is called *mitosis*.

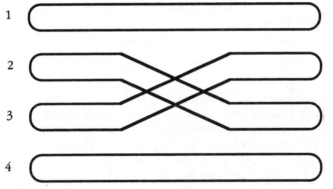

4-5   Crossover between chromosomes 2 and 3, in the foursome 1, 2, 3, 4.

What has meiosis and crossover to do with heredity and the linkage of traits?

First of all, go back to the make-believe case where height and seed shape are linked traits with genes on the same chromosome. If these are on different chromosomes, the expression of the two traits—tallness, round seed shape, dwarfism, and wrinkled shape shape—mix or assort independently of each other when passed from the first to the second generation of plants. But if the two genes are linked on the same chromosome, one would expect only two types of plants. They are shown below in the proportions they would be expected.

Tall, round seeded: 3/4
Dwarfed, wrinkled seeded: 1/4

However, in reality, matters are often not so simple, when traits are linked. Often the proportions of the two kinds of offspring are not those above. One does get some mixing of the two traits. However, the proportions are not

$$9/16, \quad 3/16, \quad 3/16, \quad 1/16$$

as would be expected if there was no linkage. Why should this be so? In fact, how could two traits mix at all in a second generation if their genes are on the some chromosome? The answer to both questions lies in crossover. Crossover occurs at meiosis in the first generation's germ cells to some extent. And in this imaginary case it might be that $2/16$ of the second generation are tall plants with wrinkled seeds and dwarfed ones with round seeds, although the previous analysis would show that plants of these kinds should not be present at all in any generation. But, assuming that the two genes are on different chromosomes, each of these two kinds of plant should together make up about $6/16$ of the second generation. However, it might be, as is often the case, that neither of these possibilities holds. How does crossover explain this?

Consider what would happen during meiosis in the germ cells of the first generation hybrid plants. Their body cells each hold a pair of chromosomes of the form

$$1 \quad T R$$
$$2 \quad d r$$

where chromosome 1 was inherited from a pure tall parent with round seeds, while 2 was inherited from a pure dwarfed parent with wrinkled seeds. During meiosis the two chromosomes pair up in the special body cells that give rise to germ cells, of which there are many. While the two chromosomes are paired up and duplicating, some such pairs can exchange their right halves to give pairs.

$$1' \quad T r$$
$$2' \quad d R$$

In these, the genes T and R (as well as d and r) are no longer together on the same chromosome, and some germ cells get one of these new chromosomes. If this is so, a germ cell having chromosome 1' above may unite with another having the same kind of chromosome. That union gives rise to a seed for a second generation plant having the chromosome pair

$$T r$$
$$T r$$

This seed gives a tall plant with wrinkled seeds. Now if the two germ cells having chromosome 2' above unite, the seed will hold the chromosome pair

d R
d R

and will yield a dwarfed plant with round seeds. Crossover would be the reason that two traits—tallness and round seed shape or dwarfism and wrinkled seed shape—do not stay together in all instances in the second generation in this make-believe example, even though their genes would be linked on the same chromosome. In reality, height and seed shape genes are not on the same chromosome. So the problem of these traits being mixed never came up in Mendel's work.

Such is not the case with most traits, and this is fortunate as far as the geneticist is concerned. The percentage of crossover between traits in a second generation of organisms has given valuable information about how genes are arranged on their chromosomes.

ARRANGEMENT OF GENES ON THE CHROMOSOME

Consider, as a general example, the dominant genes of two traits that are on the same chromosome. Let the dominant genes be A and B while their recessive forms are a and b, together on a chromosome of the same length as that containing A and B. An individual organism pure in A and B has two chromosomes of the form

A B
A B

in each of its body cells. An organism pure in the recessive traits a and b will have the chromosome pair

a b
a b

in its body cells. Such pure parent organisms in the two traits have germ cells holding the chromosomes A B and a b, respectively. Crossover has no effect in these organisms, because, at meiosis, the chromosome pairs

A B   ab
            and
A B   a b

can only exchange identical genes when attached to each other, giving new chromosomes of the same kind. So as these pure parents are crossed, the following germ cell unions take place:

$$\boxed{A\,|\,B} + \boxed{a\,|\,b} \rightarrow \boxed{Aa\,|\,Bb}$$

All first generation offspring have the traits A and B while none have the traits a and b, because A and B are dominant.

What happens when germ cells are formed in this first generation? Each first generation organism will make two kinds of germ cells, one of which has the chromosome A B and the other the chromosome a b. When two of these organisms are crossed, the possible germ cell unions are

[A|B] + [A|B] → [AA|BB]
[A|B] + [a|b] → [Aa|Bb]   Offspring having traits A and B
[a|b] + [A|B] → [Aa|Bb]

[a|b] + [a|b] → [aa|bb]   Offspring having traits a and b

if no crossover between chromosomes take place.

Generally, crossover does occur. Then, meiosis in the first generation often entails the crossover given by

$$A\ B \qquad\qquad A\ b$$
$$\rightarrow$$
$$a\ b \qquad\qquad a\ B$$

which yields two new kinds of germ cells, one of which has the chromosome A B and the other the chromosome d B. As germ cells, these are represented by the symbols [A|b] and [a|B] and unite with the other germ cells [A|B] and [a|b] in the following ways:

[A|b] + [A|B] → [AA|Bb]
[A|b] + [a|b] → [AA|bb] ---------------- Trait A accompanies b

[a|B] + [A|B] → [Aa|BB]
[a|B] + [a|b] → [aa|Bb] ---------------- Trait a accompanies B

The number of germ cells of the compositions [a|B] and [A|b] in the first generation hybrids depend on the number of crossovers between the chromosomes A B and d b that occur at germ cell formation. The more crossovers there are, the greater the number of such germ cells, made in equal numbers. But more germ cells of the forms [A|b] and [a|B] mean a greater percentage of second generation offspring having trait A along with b or trait a along with B. What determines the number of crossovers between two genes on neighboring chromosomes?

Cytologists now think that two neighboring chromosomes about to undergo crossover between genes break in two at the same point in each. Look at the chromosome in FIG. 4-6.

The point at which the pair of chromosomes breaks is totally random. It could be at any point between A and B. It seemed logical to geneticists, cytologists, and biochemists that the further apart genes A and B are on the chromosome, the more likely it is that the break will occur between them; the closer together they are, the less likely it is that the break would occur between them. This all meant that the closer together two genes are on a chromosome, the less likely it is that crossover will

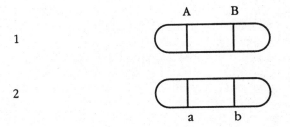

Chromosome 1 may break in two in the fashion

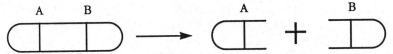

and 2 in the similar fashion

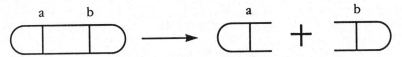

Then the short piece of 1 (holding gene A) unites with the long piece of 2 (holding gene b) as shown here

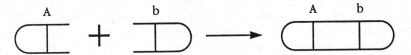

to give the new chromosome A b. Similarly, the short piece of 2 unites with the long piece of 1 as follows

to give another new chromosome d B.

4-6 The breaking and uniting of chromosomes.

occur between their dominant and recessive forms. And the further apart they are, the more likely is such crossover. But the more likely crossover is between genes, the greater will be the percentage of second generation offspring that have both trait A and trait b and trait a and trait B. And greater is the percentage of cases where the two traits separate from each other in that generation. However, the greater the number or percentage of such mixings, the further apart are genes A and B (or a and b) on their chromosome, an observation that proved very helpful to geneticists when they turned to chromosome mapping. Such mapping involved figuring out how genes are arranged on the chromosome.

The mathematics behind chromosome mapping is too complex to go into here. But the ideas behind it are easy to understand. They can be illustrated as follows: Imagine two other linked traits with genes on the same chromosome as A and B, the dominant forms of which are C and D and the recessive forms c and d. So the four traits A, B, C, and D (as well as a, b, c, and d) are linked. What is the arrangement of the four genes on their chromosome? This is a simple problem in chromosome mapping. Let's look at how the geneticist would go about solving it.

Recall that the percentage of second generation offspring in a breeding experiment with trait A along with trait b, or trait a along with trait B, gives an indication of how far apart the genes A and B (or a and b) are on the chromosome. Now what about C and its recessive form c? Does C lie between A and B or to one side of them? The same may be asked about c with respect to a and b. The question may be answered by following the traits A, B, a and b as well as A, C, a and c into the second generation in a breeding experiment, and comparing the extent of mixing of traits in that generation.

When such an experiment is done, the geneticist finds that a certain percentage of the second generation have trait A along with trait b or trait a along with trait B, and this percentage is a measure of how far apart the genes A and B (or a and b) are on the chromosome. Next he looks at the percentage of offspring in that generation which have trait A along with trait c and trait C along with trait a. If that percentage is greater than the percentage of second generation progeny having traits A, B, a and b mixed, more crossover has occurred between C and c than between B and b. But such only means that C is further from A than B on the chromosome. Likewise c is further from a than b on the neighboring chromosome. Such findings would only show that C is further from A than B. The exact arrangement of A, B, and C would not be apparent. There are two possible arrangements, as shown in FIG. 4-7.

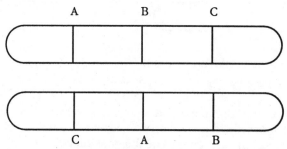

4-7 Possible arrangement of genes on the chromosome.

How can the geneticist tell which arrangement is the true one through breeding experiments? To decide this he must also look at the percentage of crossover between B and C and their recessive forms b and c. But before examining this, another interesting matter should be discussed.

DISTANCE BETWEEN GENES

At this stage, the only thing that breeding experiments could tell the geneticist was the relative distance between genes on chromosomes—that is, how many times further one gene was from a given gene than another gene—and not the actual distances between them. So geneticists made a simple, reasonable assumption: that the percentage of mixing between two traits (such as A, B, and a, b) is an exact measure of the distance between the two genes on the chromosome. However, the percentage of mixing between the two traits is a measure of the percentage of crossover between the two kinds of gene on chromosome pairs at meiosis in first generation hybrids. An equivalent way of stating the above assumption is: the percentage of mixing between two linked traits in the second generation of a breeding experiment is a direct measure of the distance between their genes on the chromosome. What geneticists did was simple. They used the percentage of mixing between two forms of two linked traits in the second generation as a measure of the absolute distance between their genes.

Let's apply this assumption to genes A, B, and C. Say that a geneticist finds that one percent of the second generation have either trait A along with trait b or trait a along with trait B. He will then say that genes A and B as well as a and b are one unit apart on the chromosome. Now say he also finds that two percent of the same generation have either trait A along with trait c or trait a along with trait C. He concludes that genes A and C (or genes a and c) are two units apart on the chromosome. He, in other words, has defined a unit of distance between genes on the chromosome in such a way that p percent of crossover between two genes means that the two genes are p units apart on the chromosome. This indicates that the three genes may have either of the arrangements shown in FIG. 4-8.

In order to decide which of these is the true arrangement, it is necessary to look at the percentage of mixing between the linked traits, B, C, b, and c in the second generation. If the correct arrangement is 1, this percentage will be about one percent. If 2 is correct, this percentage will be three percent. So the geneticist simply looks at the percentage of mixing between the A and B traits in the second generation. In the same way, he can determine the location of gene D (or d). Linked traits and the genes associated with them make chromosome mapping possible.

Therefore, the geneticist's ability to map the chromosome rests on three items:

1. The existence of linked genes on the same chromosome.
2. The crossover between linked genes on two neighboring chromosomes.

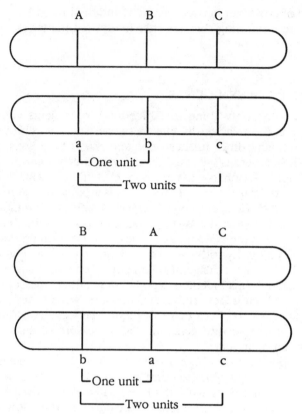

4-8 The unit distance between three genes on the chromosome.

3. The percentage of crossover between two genes on neighboring chromosomes.

Chromosome mapping has been explained in broad terms to give the general idea of the technique. The traits represented by A, B, C and D (and their recessive forms a, b, c and d) have not been given. If all traits behaved independently, as in Mendel's experiments, the arrangement of genes on the chromosome could not have been figured out. The same would be true if crossover between chromosomes at meiosis did not occur. Then the geneticist would only be able to tell that a group of linked traits had genes on the same chromosome. He would find it very difficult to decipher their arrangement on that chromosome.

CHROMOSOME MAPPING EXPERIMENTS

The three requirements above are the bare minimum needed to figure out the genetic structure of the chromosome. In real chromosome mapping experiments, many other practical considerations enter the picture. First, one needs an organism with a lot of easily recognized traits, many of which are linked and follow the simple pattern of having a dominant and recessive form. Second, the organism must have a short reproduction time—the amount of time between conception and the emergence of the new-born offspring. Humans, for example, would be an unwise choice. Their reproduction time is far too long. Besides, the number of traits in man, though large, come in many linked groups with many complex interactions between genes, many of which do not follow the simple scheme of having a dominant and recessive form. Also they are not easily discerned in the individual. Mendel's pea plants were more suited for study in this way, although their reproduction time was a little long. It had taken Mendel many years to arrive at his findings. A third characteristic of a good organism for breeding experiments is that it must reproduce in large numbers. Thus the geneticist may get a fair estimate of the percentage of crossover between linked traits. In this regard too, humans are a poor choice, while Mendel's peas plants are better. Fourth, the food (or nutritional) requirements of the organism should be as simple as possible, since, if it is to reproduce in large numbers, much food will be needed to sustain the offspring. Fifth, the smaller the size of the organism, the less space the large herds of offspring will take up, making the experimental procedure more convenient for the geneticist to handle and thus reducing his chances of error.

Clearly these five requirements eliminate most organisms from being fit candidates for chromosome mapping experiments. Geneticists found only a few suitable ones for studies. They included several one-celled creatures, the small fruit fly Drosophila, and certain strains of molds and wheat plants.

How can we be sure that what geneticists discovered about chromosomes in this limited group of organisms holds for all living organisms including humans?

Scientists have observed the chromosomes of body cells and germ cells of all kinds of organisms under their microscopes, and these chromosomes have similar physical structure in all organisms. But their length and characteristic number vary from organism to organism. They come in pairs in all living creatures. Also the chromosomes of each pair have the same length, while those of different pairs have different lengths. Take humans chromosomes as an example. Each human body cell holds 46 chromosomes, coming in 23 pairs. One chromosome of each pair was inherited from the individual's mother and the other from the father. These comments also apply to all organisms that have sex. The two chromosomes in each pair are known as *homogeneous chromosomes* because each has the same number of genes, which belong to a certain linkage group of traits, though the genes on one member of the pair may be for different aspects of the traits (like tallness and dwarfism in plants) than the corresponding ones on the other member of the pair. The little fruit fly Drosophila has eight chromosomes that come in four pairs of various lengths. Pea plants have 14 chromosomes in seven pairs. Microscopic studies have revealed the same pattern for all organisms that reproduce sexually; all their body cells hold a certain number of pairs of homogeneous chromosomes of different lengths. One chromosome of each pair is passed into their germ cells at meiosis, giving germ cells with half the number of chromosomes as body cells. But each germ cell holds one of each kind of chromosome found in body cells. So chromosome structure and behavior are similar in all organisms studied—both those suited to mapping experiments and those that are not. That is an important observation. It gives some credibility to the conclusion that facts learned about chromosome behavior in a restricted number of organisms pertain to all organisms.

Secondly, there are many small groups of linked traits in humans and other large organisms that can be followed from generation to generation. Many of these genes follow the simple patterns of dominance and recessiveness. When studied, they conform to the same laws governing chromosome structure in one-cell creatures, fruit flies, and pea plants.

The first intensive study of chromosome mapping was done by T. H. Morgan of Columbia University in 1910. It had become clear that even the simplest organisms have too many traits for the chromosome to be the same as the gene. Even one-cell creatures with a single chromosome could easily have over 100 discernible traits, so the chromosome would have to have at least 100 genes. Morgan suspected that the passing of traits from parent to offspring independent of one another, as in Mendel's studies, was far too simple a picture of the real process of heredity. Also, much remained to be learned about inheritance and how genes are arranged on the chromosome. Which genes are on which chromosomes? Are genes lined up after another along a chromosome—that is, are they in linear order—or are many of them on top of one another throughout the volume of the chromosome? These questions bothered many geneticists in Morgan's day. He, however, was a clever, hard working investigator who set out to answer them by making use of two prime tools of genetic research: the microscope and the well planned breeding experiment. The candidate for study in his experiments was the small fruit fly Drosophila.

The little fruit fly neatly satisfied the three main requirements of a good breeding experiment as well as the five additional ones relating to the practicality of the experiment. The fly has four groups of linked traits, and exhibits much crossover between genes. The percentage of various crossovers are fixed and regular. Also, many body traits of the fly are easily singled out or recognized using a magnifying glass or lens. Its reproduction time is about 12 days. It can be reproduced in large numbers; two mating flies give thousands of offsprings. Its food requirements are simple and inexpensive. Also it is a small organism, so that breeding it in large numbers causes no inconvenience or expense at all.

MUTATIONS

Morgan realized the value of the little fly for chromosome mapping research and got to work with it. His work shed much light on another interesting genetic phenomenon: that of mutation, which can easily be explained by an example. The normal eye color of the flies Morgan worked with was red. But in one of his early experiments, a strange occurrence was noted. Morgan had been satisfied that all his flies were pure in the trait of red eye color. After all, he had never gotten any flies with another eye color no matter how many times he crossed them. But finally he noticed a fly with white eyes in one of his breeding experiments. This was the first time he had seen white eye color in Drosophila. How could it be explained? Both parents of the white-eyed fly were supposed to be pure in the trait of red eye color. So the gene for white eyes could not have been inherited. It must have come into being somehow in the parent's germ cells prior to the odd fly's birth. Now if the new gene for white eye color had existed in the ancestors of the fly other cases of the trait should have turned up in earlier experiments. But such was not the case. Morgan was absolutely sure these flies were pure in the trait of red eye color. Such sudden appearance of white eye color was an example of a *mutation*. A mutation is any change or modification in the structure of a gene that gives rise to a new aspect of the trait the gene controls, or which destroys the trait so that it is absent in the organism.

Other gene mutations were known. They became of much interest to biologists and geneticists because they made the process of biological evolution better understood. One common belief before Morgan's time held that new traits were in some way acquired by the organism and passed to its descendants. The discovery of mutations was beginning to disprove this theory. The quick emergence of white eye color in Drosophila was an example. The new trait just sprung up spontaneously. It was not acquired because of any kind of effort on the fly's part. It was shown that the evolution process happened not because of any self-acquired traits in organisms, but because of mutations of traits over many years. But many of the traits an organism may acquire through mutation are not beneficial, and leave it less able to cope with its environment, so that the individual having the new trait dies out in one way or another. But some mutations give new traits which enable the organism to better cope with a new environment. Individuals endowed with these survive. This minority of beneficial mutations is the drive behind evolution, which came to be seen as "the survival of the fittest."

This was one reason Morgan had a keen interest in mutation, though his work with fruit flies was attractive to him for other purposes. One, of course, was the light the research shed upon the workings of inheritance, in particular on whether or not traits were always passed to offspring independent of each other as Mendel believed.

The white-eyed fly was a male fly. Morgan took advantage of it and let it mate with one of the normal red-eyed female ones. From the cross came 1237 offspring; all had red eyes, which clearly showed that white eye color is recessive, while normal red eye color is dominant. Such is often observed. Most traits occurring because of mutation are recessive. Thus the first generation of hybrid flies were all red eyed. But what would happen if these first generation flies were crossed with each other? What eye colors would the second generation flies display? The answer to the second question should be no puzzle at this stage, and the second generation flies should have both red and white eyes, but that is not what bothered Morgan and his coworkers, who were more interested in the relative numbers of red and white eyed flies coming out of each cross. Their real aim was to rigorously test Mendel's idea that traits are inherited independent of one another. When it comes to this question, only the proportions of different offspring in the first and second generations are of significance. Morgan, in fact, followed two linked traits into the second generation in this example: eye color and sex. Eye color came in two forms, red and white, while sex was either male or female, with both traits being linked in Drosophila.

Let's see what Morgan found when he followed eye color into the second generation. A typical result of the mating of two first generation flies to give a second generation in the above example would give, say, 3470 red-eyed flies and 782 white-eyed ones, which, on the average, was in accord with Mendel's theories about the passing of a single trait into that generation. But Morgan found something in total contradiction to Mendel's law of independent assortment: all flies in the second generation that were female had red eyes, while the males in that generation had both red and white eyes. White eye color and male sex stayed together in the second generation. Clearly the two traits were linked. Eye color and sex were totally linked in the fly.

Morgan found many other examples of linked traits in the fly having genes on the same chromosome; that was the real breakthrough of his research. In 1910, many ideas about the relationship between chromosomes and heredity were purely speculation, though it was apparent that chromosomes seen under the microscope behaved in many ways like the genes envisioned by Mendel. But real experimental proof was lacking because an organism (other than the pea plant) that was well suited to genetic experimentation had not been found until Morgan and others started working with fruit flies and one-celled organisms. Then better breeding experiments could be done and the picture brightened. Geneticists could then compare theory with experimental results and sort out good theories from poor ones. That is just where the little fruit fly Drosophila came in handy in 1910. Because large numbers of offspring could be obtained through its use, percentages of crossover could be measured with greater accuracy and the locations and distances between genes on the four chromosomes of the fly could be determined better than in other organisms. Much of our knowledge of chromosome structure and the physical basis of heredity is owed to the little fly.

IMPORTANT CONCLUSIONS

Three items about the operation of heredity came out of Morgan's work. First, the observation of Drosophila's chromosomes under the microscope and breeding experiments showed that chromosomes are the carriers of genes before Avery and his colleagues showed DNA is the genetic material. Secondly, the research conclusively showed that linked traits belong to genes on the same chromosome. Third, and quite important, the mapping of the fly's chromosomes demonstrated that genes are arranged linearly along the chromosome, with each gene making up a small segment of a chromosome. The segments are lined up one after another along the length of the chromosome.

Around 1940, the question of the relationship of the gene to the chromosome was fairly settled, and our knowledge in this area could be summarized as follows: Each organism has a fixed number of chromosomes in its cell nuclei which is the same for all members of the species. For humans that number is 46. For the fruit fly Drosophila it is eight. For pea plants it is 14, while some single cell organisms have only one chromosome. Chromosomes come in pairs. Each pair holds two chromosomes of the same length in most cases having the same kinds and numbers of genes, although those on one chromosome of the pair may be for different aspects of the traits than those on the other chromosome. The important point is that each chromosome holds genes for a set of linked traits, while one chromosome of each pair is inherited from the organism's female parent and the other from its male parent. Genes are not distributed through the chromosome, but are arranged one after the other along its length, each making up a small segment of the chromosome. By genetic mapping geneticists were able to pinpoint these segments and could estimate how much of the chromosome each occupied. Finally, the nature of the gene mutation process was uncovered. More will be said about that later.

So the first step in discovering what Mendel's genes are was taken early in this century. The physical nature of genes has been established. The gene is not the chromosome seen in the cell nucleus. Instead, it is a small segment or portion of it.

Chapter **5**

The fine structure of the gene

*T*he long period starting with the discovery of Mendel's work at the turn of the century and ending with the Meselson-Stahl experiment in 1958 may be called the first era of modern genetics. It began when Mendel postulated the gene as the carrier of traits. He thought of it as a small, physical factor in the cells of organisms. He was sure of its physical or particle nature, though he knew little more than that. Because efficient microscopes and dyes for coloring cell contents had not been developed, he could only conclude that the gene was a physical body of some kind, a belief that was supported by the fact that recessive traits like dwarfism, green seed color, and wrinkled seed shape in pea plants came through unblemished in the second generation after being absent in the first. The only idea that could explain this was that the carriers of these traits stayed intact in cells of the first generation hybrids as particle-like entities that are passed into seeds of offspring from generation to generation. No other explanation seemed to work.

PUTTING THE PIECES TOGETHER

Bigger and better microscopes were developed by the turn of the century. Also certain coloring agents that made chromosomes and other cell organelles visible were discovered in Germany in the 1870s and 1880s, while chromosomes had been seen in cell nuclei of many different species. Both cell division (or mitosis) and germ cell formation (meiosis) were revealed under the microscope. It was observed directly that half of the cell chromosomes after duplication were given to each daughter cell in mitosis. Such was similar to the way a parent organism was believed to give half of its genes to the first cell of the new-born organism. So a parallel was seen between chromosomal behavior and that of genes in Mendel's theory. Thus arose the belief that either the gene was the chromosome, or at least, the chromosome might be the carrier of genes. Other features of chromosomal behavior also matched those of genes. It became clear that the chromosome and the gene were very related in any case.

Then came Morgan's proof that many traits are linked in that they tend to be inherited together in offspring. Around the same time, crossover between genes on the same chromosome set was discovered under the microscope, and this, along with linked traits, showed that the gene thought of by Mendel did indeed exist as a physical body in cells, and was, in fact, a segment of some chromosome. These facts were well established in 1935. But they were now based on microscopic evidence and rigid experimentation.

However, the problem of the nature of the gene was being attacked in another way: that of chemical analysis in the test tube. This approach really began with Griffith's work in 1928. The conclusion of all such work was that some substance was absorbed by one-celled organisms that caused a change in certain traits they possessed, after picking up the substance. This substance was shown to be located in the chromatin of the cell nucleus. Such conclusions were the same as those reached through breeding experiments and looking at cells under the microscope, so that in the 1940s the idea that the gene was a small segment of the chromosome was well founded. Yet the chemical side of the question was not so clear cut. Chromatin, the material of chromosomes, had been shown to be composed of two substances, proteins and nucleic acids. The puzzle had centered around which substance was responsible for the action of the gene. The first theory was that only the protein portion could serve as the genetic material, because it was believed that only a protein molecule could build another molecule like itself, which brings us to Avery's experiments in 1944. They showed, contrary to current beliefs, that the nucleic acid part of the chromosome was the genetic site while the protein functioned in some secondary role. Then came the Watson-Crick model of the DNA molecule in 1953. Several years later the Meselson-Stahl experiment proved beyond reasonable doubt that DNA was the genetic material.

So two facts had been discovered in the period from 1900 to 1960. First, the question of what the gene is had been answered on the level of cytology, the study of cell structure and function, with each gene coming to be seen as a small part of some chromosome. Second, the chemical composition of these segments, or genes, was figured out, with each being made of DNA and protein, while, in addition, it was shown that DNA was responsible for the action of each gene; this was in the 1940s and 1950s. The conclusion was inevitable: the gene must be some portion of the DNA molecule. But what kind of portion? For instance, how many nucleotide pairs did a given gene entail? Most important, how did the nucleotides of the DNA chain contain the instructions, or chemical blueprint, for the creation of all the peptide chains needed by the cell, both enzymes and structural proteins? How did these instructions get from the nucleus, where the genetic material (DNA) stayed all the time, to the ribosomes of the cytoplasm, the sites of protein synthesis? How were they carried out when they got there?

These were just some of the questions biochemists pondered in the late fifties and early sixties. They will be a major topic through the rest of this book. To answer them, biochemists had to probe the gene. But so far only the chromosome had been probed in order to learn where the gene was located in the cell and what it was in terms of cell structure. Now that the gene had been shown to be a small segment of the chromosome, attention was turned to its molecular nature. So while the chromosome was probed for an inner structure in earlier experiments such as Morgan's with fruit flies, the gene would now be probed in the same way.

Around this time genetic mutation became a powerful experimental tool. If biochemists could get at the molecular basis of genetic mutation, which, on the surface, appeared as some change in the trait a gene controlled, they believed they could learn a lot about the chemistry of the gene and its finer structure. It was in experiments in this area that viruses and one-cell organisms were useful in genetic research. These smaller living creatures were much better than Morgan's fruit flies in this regard. They could be grown in larger numbers and had a reproduction time much shorter than bigger organisms. Their food supply was simple and inexpensive. Usually, one-cell organisms could be grown in a water solution containing glucose and a few salts of nitrogen, while, as we shall see, these small living bodies were the center of much genetic research in the sixties and seventies. But now let's return to mutation and how it helped biochemists probe the gene.

MUTATIONS IN THE LABORATORY

Precisely, what is the significance of genetic mutation for genetic experimentation? The answer goes back to Morgan's work with Drosophila. Mendel was lucky in one important way: he had an organism with easily recognized traits, each of which came in two distinct forms that could be manipulated and followed through two generations or more. But matters were not so simple with fruit flies. Many of its traits, such as eye color, came in one form only. Normal eye color in the fly is red, and both genes of the pair for eye color are normally for red eye color. The same is true for many (though not all) of the fly's traits. If geneticists could not get around this, the value of the fly for research would have suffered greatly, because nothing could be learned by following a trait having only one expression between generations. If, for example, plant height came only as tallness, seed color only as yellow, and seed shape only as roundness, Mendel could have discovered nothing about heredity and how it operates. There would have been no contrasting features to follow from parent to offspring. All plants in all generations would be tall with round, yellow seeds. No information about heredity and how it operates could be gotten under these circumstances. An investigator would not be able to see any laws in its operation, while the existence of the gene as a small, physical body would never have been suspected. However, not all the fruit fly's traits followed that pattern, though a lot of obvious ones like eye color did, which could greatly subtract from its value as a genetic tool if this situation could not be gotten around. But nature intervened to change this state of affairs in the form of mutations. The first example noted by Morgan was the sudden appearance of white eye color in one fly out of thousands. Such mutation did not happen often. But out of a large number of flies one or two did turn up occasionally, and that was all that Morgan needed, since such mutations proved to be recessive traits that could be followed alongside the normal trait through generations of flies. Somehow the gene for red eyes in one of the germ cells of the fly's parents or ancestors had undergone a change that turned it into a gene for white eyes. Once this recessive trait had shown itself, even in one individual, that individual could be crossed with a normal red-eyed fly to give many offspring with the recessive gene, which, in turn, could be crossed to give many white-eyed flies. Many other spontaneous mutations turned up among large herds of fruit flies that affected other traits such as eye shape, body color, body shape, wing shape, and wing design. Yet they were quite rare. One had to raise many thousands of flies and examine them with scrutiny to find these rare mutations. Would it not have been preferable if one could somehow step up the rate at which such mutations occurred? This idea dawned on many geneticists at the time of Morgan's work.

To step up mutation rates one would need a way to artificially cause mutations in the laboratory. How could this be done? Nobody had any clue as to how nature accomplished genetic mutation at the meiotic phase of germ cell formation. So how could anyone bring it about in the laboratory? Geneticists tried all kinds of procedures to create mutations such as exposing cells and organisms to a wide gamut of chemicals, but with no success. No one could cause a single mutation. Part of the reason was that the genetic material, chromatin, was buried deep within the cell nucleus and could not be reached by such chemicals, although, at the time of the Morgan's research, a breakthrough in this endeavor came about. The brilliant geneticist H. J. Muller happened to expose fruit flies to X-rays, an electromagnetic radiation like visible light, but more energetic. What he found in their offspring was truly amazing. Flies with many different mutations turned up among them, some of which had smaller wings than normal, or no wings at all, while others had abnormally shaped eyes, bulging eyes, and so on. But where these characteristics really mutations? There was only one way to find out: a cross between the abnormal flies and normal ones. Muller did this with many odd traits induced by X-rays and found they were passed from generation to generation in the same way as recessive traits in Mendel's experiments, showing beyond a doubt that they were true gene mutations caused by the exposure of normal genes to X-rays. X-rays are energetic and penetrating enough to get at cell nuclei and affect chromosomes. Many new and bizarre mutations were created in the fly through their use. This was a great aid to chromosome mapping.

In the following years the ability to create mutations proved very valuable to working out the chemical structure of the gene; the rest of this book centers around this story. A further breakthrough, for both theory and experiment, made use of artificially induced mutations to learn how genes are related to the enzymes they made to carry out the chemistry of the cell. An experiment done by George W. Beadle and Edward L. Tatum of Stanford University in the early 1940s was the first of a series of developments that shed light on this important question, and led to our present knowledge of the tiny chromosome segments called genes. The experiment made use of a bread mold, Neurospora, that underwent artificially induced mutations quite well and had other attributes that made it a good candidate for the experiment. An entire section is dedicated to this history-making experiment and the theory it prodded Beadle and Tatum to put forth. It will give interesting historical background about how the molecular structure of the gene was worked out, and will be an excellent introduction to the method modern biochemists and geneticists used in the sixties and seventies to get insight into the genetic code. Although the experiment was performed in the early 1940s, it was very contemporary in its approach.

THE ONE GENE-ONE ENZYME HYPOTHESIS

The Beadle-Tatum experiment was really the end result of earlier findings and speculations among biochemists. It had been shown in the 1920s that enzymes were proteins, and it soon became clear that enzymes carry out the process of heredity. Each trait comes about through some chemical or chemicals, made with the help of enzymes working in the cells of the organism; that much was known in the 1940s. But the chemical details were missing. Did many genes control the making of one enzyme, or did one gene control the making of many specific enzymes? Or was there a one-to-one correspondence between genes and enzymes, with each gene controlling the production of just one enzyme?

Actually experimental data, shedding some light on this question, was available in the nineteenth century when the English physician, A. E. Garrod, made a study of the human abnormality known as *alcaptonuria*, in which a person passes brown instead of yellow urine because of the presence of the substance *alcapton*. The disease was found to be a simple recessive trait following Mendel's laws. Garrod's work showed that most people have an enzyme in their system for the breakdown of alcapton into carbon dioxide and water, while those having the disease lack that enzyme and pass the substance into the urine. Let's look at these facts more closely.

Alcaptonuria is inherited as a simple recessive trait. So it should be governed by a mutated gene, the normal form of which controls the breakdown of alcapton into carbon dioxide and water. And since it is inherited in the same way as Mendel found recessive traits like dwarfism, green seed color, and wrinkled seed shape to be, we can be sure it is governed by a single gene. People having alcaptonuria have two genes for the disease in their cells. Normal people have at least one gene for the breakdown of alcapton. That gene is dominant and most people have two genes of this kind in their cells. Garrod's research implied that the normal gene controlled the making of the enzyme which helped break down alcapton; and here was clear evidence that one gene directed the production of one enzyme.

Other work showed that normal eye color (red) in the fruit fly Drosophila is a result of a series of chemical reactions controlled by an enzyme made by one gene—that for red eye color—and when that gene undergoes a mutation, the structure of the enzyme molecule which regulates the making of the red eye pigment is altered and is unable to make the pigment. So if both genes for red eye color are mutated in this way, the fly will have white eyes. Evidence was gathering in favor of the hypothesis that one gene controls the making of one enzyme.

All these developments inspired Beadle and Tatum to do a rigidly controlled experiment to test the one gene-one enzyme hypothesis. But, first, a word about the difference between a *theory* and *hypothesis*.

A hypothesis is an intelligent guess a scientist feels might explain certain findings of a limited scope; it is an idea that seems to make some specific observations understood. Take the one gene-one enzyme hypothesis at the time of the Beadle-Tatum experiment as an example. Among the limited findings of the time were the results of Garrod's work with alcaptonuria as well as those regarding the making of the red eye pigment for normal eye color in the fruit fly. Both findings suggested that one gene was responsible for one enzyme. But this was not enough information to conclude that the hypothesis was valid over a much wider realm, and further experimentation was needed; the one gene-one enzyme idea was only a hypothesis.

A theory, on the other hand, is a more firmly established scheme that has succeeded in explaining many different findings, often in many different areas, to an extent that makes the scientist satisfied he is on to a valid line of explanation, at least for the large spectrum of observations the theory explains. A theory is experimentally verified over some large span of observations. Also, it can often predict new findings and suggest ways to bring them to light experimentally. Mendel's gene idea was a good example of a theory at the turn of the century. The Beadle-Tatum experiment was a major event that helped raise the status of the one gene-one enzyme hypothesis to that of a theory.

The organism used by Beadle and Tatum was the bread mold, *Neurospora Crassa*, having a very neat characteristic that made it well suited to their experiment: it reproduced by means of small bodies called spores that came in packages (or sacs) of eight apiece. The spore grows into the mold when placed in a nutritive medium which contains all the mold's essential nutrients. Among these are carbohydrates, amino acids, and vitamins. The spores could be removed from their sac, using a microscope and small needle-like instruments, and placed in the medium of nutrients. But before this was done, Beadle and Tatum exposed the spores in the sacs to X-rays to induce mutations in the genes they contained.

The normal variety of the mold had spores that would grow on what is called a minimal medium, or one that holds the smallest number of simple carbohydrates, amino acids, and vitamins needed by the spores for growth and reproduction. The mold, however, needed many more essential chemicals as part of its life chemistry to function. But the additional ones are made from the essential ones of the minimal medium by the normal form of the mold, through enzyme controlled reactions. So spores from the normal mold could grow when placed in a minimal nutritive medium.

Beadle and Tatum knew the basic carbohydrates, amino acids, and vitamins the normal mold needed from a minimal medium, and could easily prepare such a medium. But their experiment made use of many different media. One of these contained all the simple nutrients needed by the normal mold plus all those it made from these necessary ones; this medium could be called a maximal medium.

The first thing the experimenters did was to expose many spore sacs from wild, normal Neurospora to X-rays to create possible mutations in them, since they wanted to know how such mutations affected the nutritional needs of the mold. Many of the irradiated spores were placed on a maximal medium and gave rise to many different colonies of the mold. Next Beadle and Tatum took spores from each of these colonies and placed them on a minimal medium. If a given kind of spore grew on this medium, it had not suffered any mutation that affected the mold's ability to make any essential carbohydrate, amino acid, or vitamin from the basic ones of the minimal medium. But spores from many colonies that grew on the maximal medium did not grow on the minimal one. These had no doubt suffered one or more mutations that rendered the mold unable to make one or more of its essential life chemicals from the simple ones in a minimal medium. What particular substance, or substances, was such a mutant mold not able to make?

To answer that question, the experimenters made many other nutritional media. Each of them continued all the basic carbohydrates, amino acids, and vitamins of the minimal medium plus just one variety of the more complicated substances normally made by the cells of the mold. There were three types of such media. One contained all the simple nutrients of the minimal medium plus all the more complicated carbohydrates normally made by the mold from the basic ones. Call a medium of this type medium 1. Another type held all nutrients of the minimal medium plus all additional amino acids made by the mold. Call this kind of medium medium 2. A third type held all the nutrients of the minimal medium plus all the vitamins needed by the mold. Call this type medium 3.

Next, the mutant spores that would grow on the maximal medium but not on the minimal one were placed on media of types 1, 2, and 3. If a certain spore grew on medium 1, it lacked the ability to make an essential carbohydrate from the simple ones of the minimal medium. Thus the X-ray treatment had apparently caused a mutation that interfered with the mold's ability to make one or more of its essential carbohydrates from the simpler nutrients of the minimal medium. If another type of spore grew on medium 2, it lacked the ability to make one or more amino acids from the basic nutrients in the minimal medium. The X-rays had apparently produced a mutation that affected the mold's ability to make amino acids. Likewise if the spores from a colony that had not grown on the minimal medium grew on one of type 3, it lacked the ability to make one or more essential vitamins from the nutrients of the minimal medium; the spore had suffered a mutation that affected the mold's ability to make vitamins.

Beadle and Tatum had isolated many mutant colonies of Neurospora that grew on each of the three media. But before they could learn anything about the relationship of genes to the enzymes that helped make the life chemicals of the mold, they needed to go further yet. For example, did a mutant of type 1 lack the ability to make just one essential carbohydrate or many? Did one of type 2 lack the ability to make one essential amino acid or many? Likewise, did one of type 3 lack the ability to make just one essential vitamin or many? To find answers, the investigators first took many mutant spores of type 1, and tried to grow them on many different media. Each of these media held all the essential nutrients of the minimal medium, plus one additional carbohydrate that is normally made in the cells of the mold. If such a mutant spore grew, for example, on a medium in which the added carbohydrate was more complicated than glucose, the mutant mold could not make such an essential sugar from the nutrients of the minimal medium; the sugar had to be supplied in the medium in which the mold grew. Similarly, each mutant of type 2 were placed on different media, each containing all the basic nutrients of the minimal one plus one specific amino acid. In this way many mutant forms of Neurospora that could not make a given amino acid that the normal mold could make were found. Also, in the same fashion, many mutants were found that could not make a specific essential vitamin that the normal mold could make from the basic chemicals of the minimal medium. In other words, many mutations were studied that wiped out the mold's ability to make just one needed carbohydrate, amino acid, or vitamin from the simple ones of the minimal medium. The essential chemical had to be supplied in the growing medium.

Beadle and Tatum also looked at the way this inability to make an essential chemical was passed on to future generations of the mold. They found it was passed on in the same fashion that simple recessive traits in

Mendel's experiments were passed on, which showed each such trait was governed by a single mutated gene. This gene was always recessive. All this could only mean that the normal gene controlled the making of the essential chemical.

Now it was known that the production of each essential chemical in the organism was under the control of a specific enzyme. It thus seemed that a mutation somehow blocked the production of an essential enzyme. From their data, Beadle and Tatum advanced the one gene-one enzyme theory, which held that each gene was responsible for the making of each enzyme needed by the cell. The results of their experiment (and others) seemed to support this theory, though eventually some notable exceptions turned up.

Sometimes a mutation of a single gene leads to an organism's inability to make two of its essential substances. Again this trait seemed to be governed by a single mutated gene, as indicated by the way it was passed to offspring. First, it was believed that such a gene controlled two different enzymes, and the one gene-one enzyme theory would then be wrong. Also there were cases where a mutation in two different genes led to an inability to make one essential substance needed by the organism; it was as if the making of one enzyme sometimes required two genes. But further research showed that these observations did not discredit the main idea behind the one gene-one enzyme theory, since when two enzymes seemed to be governed by the same gene it was discovered that one enzyme controlled the production of both of the substances previously thought to each require a separate enzyme. A mutation in the gene controlling this enzyme would destroy the organism's ability to make both substances. Also, when the making of one enzyme seemed to be under the control of two genes, further study showed that two enzymes were involved. The enzyme E_1 regulated by one of the two genes helped make one of the essential substances—call it B—from one of the minimal medium—call it A—while the enzyme E_2 regulated by the other gene helped convert B to another essential substance C, or in symbols,

$$A \xrightarrow{\quad E_1 \quad} B \xrightarrow{\quad E_2 \quad} C$$

Thus, there are really two enzymes, each controlled by one of the two genes, once again saving the one gene-one enzyme theory.

MODIFICATIONS TO THE THEORY

Not all findings were so favorable to the theory. The scheme had to be modified in some cases, though not drastically. The modification came through a study of two well known blood diseases resulting from irregularities in the hemoglobin molecule: one amino acid unit in one of the peptide chains of the molecule is different from what it is in normal hemoglobin. Genetic studies showed that both diseases were inherited independently of one another, or were separate traits, so different genes had to be responsible for them. But hemoglobin was known to be a single protein that should be under the control of a single gene by the one gene-one enzyme theory. Each disease should have come about through a mutation in that gene, which normally regulated the making of the usual hemoglobin, although the two diseases were inherited independently. So two different genes did seem to be involved in making the hemoglobin molecule. Further research proved such to be the case.

Consequently, no doubt remained that the one gene-one enzyme theory had to be modified. The catch was that the hemoglobin molecule is made up about 600 amino acid units arranged in four completed peptide chains. Chemical analysis showed that the amino acid substitutions of each disease were on different peptide chains in the molecule. This suggested an explanation. What if each of the four chains of the large molecule were constructed under the direction of a different enzyme? Then each chain would be under the control of a different gene. The two diseases could be inherited independently. In this way the one gene-one peptide chain theory came into being. This change was not drastic. It only covered cases where the simpler one gene-one enzyme theory did not seem to work, although many enzymes had only one peptide chain and the theory in its original form held for them, thus showing that Beadle and Tatum did discover something very significant, though more than one gene is sometimes involved in making more complicated proteins. So far no exceptions to the one gene-one peptide chain theory have been found. But the gene concept gets quite nebulous as we approach the molecular level in the study of life.

From the example with blood diseases, it is evident that a mutation leads to some change in molecular structure in the enzyme a gene controls (or partially controls). Remember that DNA in chromosomes consists of chains of deoxyribose nucleotides. But each gene was thought to be responsible for a given peptide chain. It therefore followed that a mutation in some way changes the molecular structure of the portion of the DNA chain constituting the normal gene. Also, the mutated form of the gene would lead to a series of chemical reactions that would make a peptide chain of a different amino acid composition from the normal peptide chain of the gene.

But what exactly was the nature of this change in the DNA structure of a gene known as a mutation? Could there be more than one mutation in a given gene? If the answer to the last question was yes, the biochemist and geneticist could have a means to probe the inner structure of the gene, which had been established to be a small portion of a chromosome in the cell nucleus. It turns out that the answer is yes. This led finally to an understanding of the exact molecular structure of the small bodies called genes. I say "called" because such discoveries brought much ambiguity into what the term *gene* really means.

The finding that a single gene might undergo many different mutations came about by the use of bacterial viruses as tools of genetic research, one of which, the T4 virus, gave us much of our present knowledge of the genetic code. Let's look at how this virus was used to probe the gene.

THE T4 VIRUS AS A GENETIC TOOL

You can easily understand why viruses were good tools for probing the gene if you remember some characteristics of larger organisms like fruit flies and bread molds. One characteristic had to do with the fact that the amount of crossover between linked genes made chromosome mapping possible, because the closer together two genes are on the chromosome, the less likely it is that crossover will take place between them. That is, the closer they are, the larger the number of offspring that must be raised in a second generation to give a sufficient number that show crossover between the two genes. That in turn meant that the mapping of genes very close together on the chromosome required an organism that reproduces in very large numbers in as short a time as possible.

Prior to viruses, the best organism for chromosome mapping was the fruit fly *Drosophila*. Two mating flies give at most several thousand offspring, which was good enough to learn how genes were arranged along the chromosomes. Most of the fly's genes were far enough apart to give a sufficient number of offspring among the thousands that showed mixing of the traits of the genes, making it clear that crossover had occurred between them. But even then a strange fact had been observed. In the case of some traits displayed by the fly, one gene seemed capable of showing more than one kind of mutation. Such cases were rare, but they tended to show that what Mendel thought of as a single gene might have some kind of inner structure. This structure, of course, could not be adequately probed using fruit flies and bread molds, because the parts of a single gene would be too close together on the chromosome—so close, in fact, that the amount of crossover between them would be extremely small in any case. Many more offspring than these organisms could give would be needed to detect such crossover. So an organism that reproduced in much larger numbers was needed; one that reproduced in the millions rather than in the thousands.

Another drawback with the fruit fly for probing the finer structure of the gene was that two kinds of flies were required, male and female. This, along with the former setback of insufficient numbers of offspring, made for an added expense and bother in genetic research. But sex had another experimental disadvantage: mutations, being recessive traits, were hidden in the first generation since the dominant normal form of the gene prevented the mutations from showing in the first generation flies. Two generations of flies had to be raised before mutations would show up. That, too, was an added burden as more offspring became necessary to accommodate it alone. It would have been nice if the geneticist could work with a very small organism that did not reproduce sexually, having only one set of chromosomes holding genes for only one aspect of each trait displayed by the organism; one such organism could be found among the one-celled ones that reproduce by cell division (or mitosis). Chromosomes in these organisms do not come in pairs. They hold only one of each kind of chromosome.

Before cell division, each chromosome makes an exact copy of itself, which goes to one of the daughter cells, while each parent chromosome goes to the other daughter cell, so that both daughter cells have exactly the same kind of chromosomes as the parent cell; at least, that is what usually happens. However, sometimes chromosome duplication in the parent cell is not perfect. Occasionally a gene mutation arises in a duplicated chromosome in the parent cell. Some genes may be copied incorrectly from a mother chromosome, and such mutations are in turn passed to the offspring cells so that some daughter cells will have some traits different from the original cells. Mutations may also be induced in one-celled organisms by exposing them to X-rays or various chemicals. These simple organisms have the advantage that such mutations show up clearly in the first generations. Because of their lack of sex, their use in gene and chromosome probing experiments was limited, since, despite its added burden, sex had had the advantage of crossover occurring at meiosis. That provided much information about gene and chromosome structure. So it would have been nice if one could use a one-chromosome organism that displayed something like sexual mating.

Surprisingly, that is just where bacterial viruses proved excellent tools for investigating gene structure. Viruses are the simplest organisms known and consist of a large molecule of double-stranded DNA in most cases surrounded by a shell of amino acid units. But how did viruses provide the advantage that sex had in larger organisms—that of crossover and the information it gave? The T4 virus proved very useful in genetic studies of the late fifties and sixties.

A T4 virus has an intricate shaped and design (FIG. 5-1). It consists of a head made of about 200,000 DNA base pairs surrounded by a protein shell of complicated shape. The head is covered with a shell of amino acid units. However, the tail is more interesting. It is made of a hollow cylindrically shaped portion of amino acid units, off of which several long structures branch that resemble spider legs. The virus uses these leg-like bodies to bore open the membrane of the cell it invades. The DNA double helix in the head is then injected into the cell through the hollow portion of the cylindrical part of the tail. But how is the problem of lack of sex in the virus bypassed?

Biochemists found that when two or more viruses invade a given cell together their DNA cores sometimes meet and exchange parts. Thus crossover between DNA cores of viruses takes place.

Viruses must have traits because they have DNA, and hence genes, and also carry out life activities in cells they invade. Two important virus traits are the number of strains of bacterial cells a virus can infect and kill, and the shape of the plaque the virus leaves on an agar plate containing a culture of the bacterial cells the virus is invading. A plaque is simply a

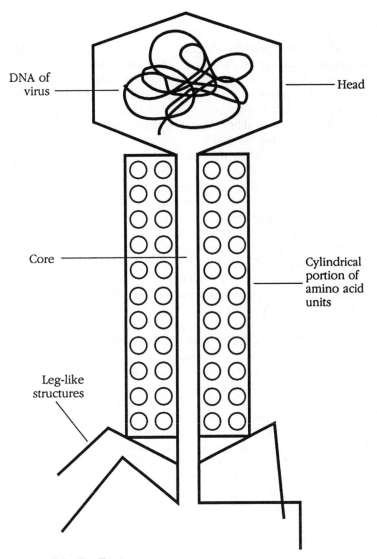

5-1 The T4 virus.

clear area of a given shape and size left on a cultured agar plate which results from the fact that many bacterial cells have been killed by the invading viruses at that place on the plate. The size and shape of a plaque left on a culture of bacterial cells after a certain time by the viruses are traits of the particular virus. A plaque grows with time, because more and more cells are infected by the viruses as time goes one. These traits, like others, are controlled by various genes in the DNA core of the virus. Another trait is the ability of the virus to kill the host cell after infection. Some viruses merely infect cells without killing them, and this is also a viral trait.

T4 viruses infect and kill *E coli* cells of two kinds known as B and K strains. A normal T4 virus put on a culture of B or K *E coli* cells on an agar plate will, after a short time, produce a visible plaque of a given shape and size, hereditary traits of the T4 virus. The single T4 virus—actually its DNA core—that first enters a given *E coli* cell of the culture makes about 100 viruses identical to itself, while inside, that upset the normal chemistry of the cell and kill it. Then 100 or more new T4 viruses are released from the dead host cell as that cell breaks apart. These go on to infect and kill other E coli cells on the plate. Then there are about 10,000 new T4 viruses which, in turn, enter other cells. After a short time, millions of new T4 viruses are produced in this way in the vicinity of the first infected cell. The small area on the agar plate consisting of all the dead *E coli* cells left after such infection is the plaque that appears as a clear area of a certain size and shape. The larger the area of this plaque at a given time after the first viruses are introduced into the culture, the more cells that have been killed by the viruses. The larger the plaque area, the larger is the number of offspring viruses made from the first one(s) brought into the culture of cells. Thus the area of this plaque gives biochemists and geneticists a means to determine the number of offspring viruses present. The larger the plaque area, the larger the number of viruses the plaque holds, and the larger the number of the first viruses that started the infection. So the size of the plaque after a certain time gives scientists a way to count viruses.

Normally a virus, while in a host cell, makes an exact copy of itself, and all the viruses in a plaque should be identical to the single parent virus introduced into the bacterial culture. But once in a while an error in the DNA duplication process may arise. This results in an offspring virus that has a mutated gene in its DNA core, altering one of its traits. Which trait is changed depends on where in the core the mutation takes place. The virus is different in this particular trait from both the parent virus and other offspring. The mutant virus can also enter a bacterial cell of the culture and make many others like itself. It is also possible that two viruses with two different mutations in different parts of their DNA cores can arise in the culture, each of which is different from the others and the parent virus in some given trait.

Imagine that two such viruses have arisen by chance in a given plaque and that each gives thousands of others like itself after a short time. Now the investigator can prepare a large number of a given mutant virus by isolating one or a few single ones. He can then put one of the mutants into a test tube of bacterial cells to get millions of the same kind through duplication. Such isolation of two different kinds of mutant T4 viruses has shown that crossover can take place between the DNA cores of the viruses in a bacterial cell being infected. Say that one of the two mutant viruses—call it 1—has a mutation in gene A—call the mutated gene a—while the other mutant virus—call it 2—has a mutation in another gene B—call the mutated gene b in this case. The DNA cores of the viruses 1 and 2 are:

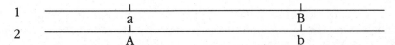

The core of 2 has the normal form of gene A, while that of 1 has the normal form of gene B. But how does crossover occur between the two cores? How can it be brought about in the laboratory?

The experimenter can arrange things so that both mutant viruses enter the same host cell, although actually only their cores enter. While inside, the two cores may come so close together that they exchange genes on relatively rare occasions. The experimenter simply takes a solution of mutant virus 1 and a solution of mutant virus 2 and mixes them to get a solution of both viruses, which can be introduced into a culture of bacterial cells. One precaution is usually taken here: the experimenter wants at least one of each kind of virus (1 and 2) to enter each host cell. So the number of viruses must significantly outnumber the host cells, or there must be at least twice as many of each kind of virus as there are host cells. Then each host cell is as likely to absorb two mutant viruses of type 1 and two of type 2. The chances are good that each host cell with be infected by at least one of each type of mutants. Both types of virus will reproduce in the same host cell. Then both kinds of DNA cores have a good chance of coming near each other in the host cells so that crossover between them is possible.

Under such conditions, two such viral DNA cores have been observed to somehow exchange parts, although biochemists are not certain about the exact mechanism of the process. It seems likely that the parts of the two cores are actually exchanged. Thus, some pairs of the two cores may go through the crossover shown below in some host cells of the culture. What is important here is that core 2″ is that of the normal parent virus which has arisen again through crossover between the two mutant cores. Biochemists found that often a few viruses like the normal one formed when two different mutant viruses infected a culture of bacterial cells; the only explanation was that some kind of crossover occurred between the cores of the two mutants while in a host cell. No other idea seemed to fit the facts.

MAPPING OF THE VIRAL CORE

But what exactly was the significance of such crossover in genetic research? It had the same relevence as sex did in research with the fruit fly *Drosophila*, and made the mapping of the viral DNA core possible, the only difference being that the virus had

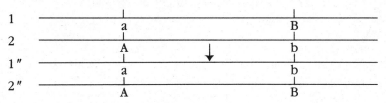

millions of offspring compared to *Drosophila's* thousands. Remember the importance of numbers in genetic experiments. The larger the number of offspring, the greater the number of highly unlikely crossovers that could turn up among them. Also the closer together two genes are on the chromosome, the less likely is crossover between them, and the fewer will be the number of offspring showing crossover between them in the population, if any show it at all. So the closer together two genes, the larger must be the number of offspring to detect crossover between them, let alone to determine the percentage of crossover, which is necessary to learn the exact location of genes. For that purpose, thousands of offspring were sufficient to map *Drosophila's* chromosomes; its genes were far enough apart. But to probe the inner structure of the gene, many more offspring were needed. Only chromosomal distances down to about the length of an average Mendelian gene could be mapped using fruit flies and larger organisms having sex.

It was in probing the fine structure of genes that the millions of offspring given by the T4 virus came in handy. Mapping of the more obvious T4 genes on its core could be done easily by letting two mutant viruses infect the same culture of bacterial cells in the way just described. Most of the offspring would be like one parent or the other, while the number of each kind of offspring could be counted by plaque size or area. The large plaques are tested intricately to see if any of the normal T4 viruses are present, and if some are, they could only arise through crossover between the cores of the two mutant viruses. Shortly we shall see an example of how all this is done and how it made modern history in genetics.

The few normal viruses are then isolated from the rest and put into a culture of host cells on an agar plate. They quickly form a plaque on the plate. The area of the plaque after a given time will show how many of the normal viruses were originally present. If one was present, the plaque will be a certain size after the given time, while if two were originally present, the area will be twice what it would be if one was originally present, after the given time passes. If three were originally present, it would be three times as large, and so forth. So the biochemist or geneticist can figure out the exact number of normal T4 viruses coming out of the cross using plaque size as an indicator. From this and a knowledge of the total number of offspring, he can arrive at the percentage of crossover between the two viral DNA cores with great accuracy, and by doing this with many different mutant T4 viruses, he can map the viral DNA core, the crucial point being that the large number of offspring makes possible an estimate of the percentage of crossover, which is very small. The experimental procedure using T4 is also simpler than others, since the virus has only one set of genes. Thus mutations are not blocked from expressing themselves by the normal genes.

Mutations in fruit flies and larger organisms turned out to be simpler than many in the T4 and other viruses. In larger organisms, mutations were simply a few alterations or variations of simple easily recognized traits such as eye color and wing shape. Those arising in T4 viruses seemed to be many distinct mutations of what was, up to that time, looked upon as a single gene. One such group of T4 mutants had to do with plaques the virus formed on B and K cultures of *E coli* cells. The normal strain of T4 produces a small, clear plaque on both cultures that has rugged edges. But a class of T4 mutants were found which did not give a plaque on K cultures, but gave a large plaque on B cultures with smooth edges. There are many such mutants, each of which can be identified by the shape of the plaque it leaves on a B culture. None of them leave a plaque on K cultures. Did all these T4 have mutations in a single gene or in different genes? The ability of normal (or wild) T4 viruses to leave plaques on both B and K cultures is a given trait in Mendel's scheme that should be controlled by a single gene on the DNA core of the virus, so that a T4 mutant's inability to leave plaques on K cultures is clearly a mutation of this gene. But there are many such mutants, each leaving its own shape of plaque on a B culture of *E coli*. So one gene seemed to undergo many different mutations in these small organisms.

Genetic mapping experiments had traced the gene for normal plaque formation on both cultures to a segment of the T4 DNA core known as the *rII region*. The length of this region seemed to be on the order of that of a typical Mendelian gene. It was discovered that the region held at most two genes. Yet there were many rII mutants. So at least it was established that many of the mutations were in one gene or the other of the two, and hence, in the same gene.

BENZER'S RESEARCH

The work of Seymour Benzer of Purdue University in the late 1950s showed that there were many rII mutants, occurring at many different sites in the rII region. Benzer isolated many rII mutant among T4 viruses. Each rII mutant gave a plaque of a unique shape on a B culture of *E coli* cells, and none on K cultures. These mutants were used by Benzer to map the rII region.

He grew a large number of each of two rII mutants and let both infect a given *E coli* culture at the same time, with the number of each mutant virus far outnumbering the *E coli* cells of the culture, so that each bacterial cell would have a good chance of being infected by both mutants. Then crossover could occur between the two rII mutant cores in the host cells. Benzer found that most of the offspring were like one or the other of the two rII mutants. But once in a great while a normal T4 virus would arise through crossover in his experiments. Benzer and his colleagues could then count the number of normal viruses in a B culture by the size of the plaque. He could then figure out how many normal ones must have risen through crossover, and from this and a knowledge of the total number of offspring (also determined from plaque size), he could get at the percentage of crossover between pairs of many different T4 mutants. In this way, he was able to map out several hundred points in the rII region, a big step in determining the inner structure of a gene. But the power of the T4 virus as a genetic tool went much further than this one experiment.

Plaques of T4 held such large numbers of the virus that rare cross-overs between points very close together on the DNA core could be detected. The rII region was about as long as a typical Mendelian gene. Benzer mapped many areas of the rII region in detail. In fact, he showed that the sites were often about two nucleotides apart. Because of the viruses' ability to pinpoint sites so close together on the DNA chain, it became very important in learning about the chemical blueprint for heredity held by the DNA molecule, while much of our knowledge of the genetic code came through the T4 virus. But what did the virus show about the nature of this code?

FIRST CLUES TO THE GENETIC CODE

In the late 1950s, F. H. C. Crick, the same man who, along with Watson, developed the Watson-Crick model, made several additional discoveries using the T4 virus. These had great bearing on the blueprint for heredity contained in the DNA molecule. Dyes known as *acridines* could step up the mutation rate in T4 and one-cell organisms. Crick used these dyes in his work with T4 at Cambridge University in England.

If two mutants viruses are allowed to infect the same cell, crossover takes place between some of their DNA cores. In this way, a few of the normal T4 viruses arise. The crossover was given by the diagram:

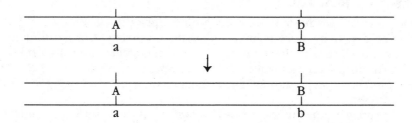

The member of the bottom pair having the genes A and B is the normal T4 virus. But the other member, with the genes a and b, is a T4 virus with both mutations. So whenever such crossover takes place, one gets not only a normal T4 virus, but also a doubly mutant one having both mutations in its DNA core. Thus an experimenter could make a doubly mutant T4 virus from two singly mutant ones. Then a whole culture of the doubly mutant ones could be made in a proper *E coli* culture. Such doubly mutant T4 could then be used to infect other *E coli* cells along with a third mutant T4 virus having the DNA core:

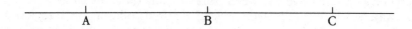

Also, a few instances of the crossover

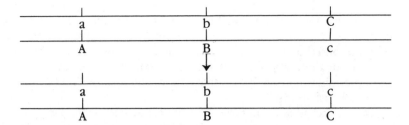

could take place, giving rise to the triply mutant virus:

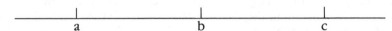

This virus has three different mutations. In the same way T4 viruses having four, five, six, seven, and so on, different mutations could be made.

Crick dealt with doubly and triply mutant T4 viruses, though he hit upon another important finding about T4 mutations: some mutations leave the gene they affect partly working, while others totally destroy the function of the gene. Crick thought this might be explained in terms of the structure of the DNA molecule.

All DNA has a molecule of two interwinding, polynucleotide chains, having nucleotides each holding one of the four bases A, G, T, or C, while the huge molecule of the T4 core was found to hold 200,000 nucleotides in each chain, and hence, the same number of base pairs. Also, DNA is the genetic material of all organisms from viruses to humans. Through certain discoveries about the chemistry of heredity, it became clear that the chemical processes involved in carrying the heredity message from the DNA chains to the protein chains was much the same in all organisms, be they viruses, cells, or humans. But the regulation of these process had to be in the DNA molecule, the molecule of heredity.

What decided the protein chains that the DNA chains would give for a given gene? What was there about the DNA chain of a given gene that determined the peptide chain the gene controls? The sugar-phosphate backbones of both DNA chains are the same throughout each; they are not capable of any variation. Only the base groups change along the polynucleotide chain, while their order along the chain decides the kind of DNA, and gives different DNAs their uniqueness. Somehow this order had to be what decided the peptide chain controlled by a given gene. But how? Let's look at the speculations of Crick and others.

CRICK'S HYPOTHESIS AND EXPERIMENTS

If it is assumed that each peptide chain is decided by the order of the base groups in its gene, we seem to immediately run into a dead end. The simplest assumption would be that each base group corresponds to a given amino acid to be incorporated into the peptide chain the gene controls, though this seems impossible. There are only four base groups, but 20 amino acids. The amino acids outnumber the bases 5 to 1. Clearly, each base group would have to stand for more than one amino acid, which was highly unlikely, since the chemical machinery of the cell would have no means of knowing which of the possible amino acids a given base group would imply, and thus would be unable to assemble the peptide chains of a gene in an unambiguous way. It seemed clear, therefore, that a single base group could not code for a single amino acid group. The DNA blueprint was not that simple.

However, the genetic message had to reside with the base groups, because their arrangement along the DNA chain was the only factor that differed in various DNAs. This problem of how the base groups carried that message was the chief preoccupation of molecular biology at the time of Crick's research, the late 1950s and early 1960s. Crick played a leading role in solving this problem. How he did so centered around his work with the T4 virus. That is truly remarkable: most of our knowledge about something as complex as the genetic code has come through an organism as tiny as a virus. It showed that the belief that the code is the same in all organisms is a sound one. Of course, it should be, because the chemicals of all living cells are of the same basic molecular structure. Some of Crick's contributions to our knowledge of the genetic code will be examined shortly.

There are two types of mutations uncovered in the T4 virus, the chief tool of this research. Benzer had shown that genes of the T4 virus held many smaller parts, each capable of mutation, and T4 genes could be probed down to sites on adjacent nucleotides. Thus it was supposed that a gene mutation may consist of some kind of change in the molecular structure of a single base group, a view boosted by other findings.

Crick came upon two kinds of mutations in T4. One left a gene totally inactive. In that case the trait the mutated gene controls was absent in a virus having the mutation, the inability of rII mutants among T4 to form plaques on K cultures of *E coli* being an example of such a mutation. A second type of mutation left a T4 gene partially working, although not in the same way the gene normally worked. These observations led Crick to come up with a theory that a mutation was a change in the molecular structure of a base group in the DNA of the gene.

At this stage of his research with T4, Crick made some assumptions about the way in which the machinery of the cell reads the genetic message held in the DNA chains. One of these was that it seemed logical that the message should be read from one end of the gene's DNA chain to the other. If this was so, reasoned Crick, any change in the order of the base groups would alter the message read. Also, adding base groups or deleting them (as well as modifying the structure of a single base group) should also alter the message read by the cell. But how could Crick account for the two kinds of mutations in this way?

It seemed likely that a mutation that left a gene partially working resulted from a change in a single base group of the DNA chain comprising the gene. The message would be read correctly before and after the altered base group, giving most of the peptide chain the correct amino acid sequence. The peptide chain would have just several amino acids different from the normal peptide chain. The several different amino acids would be those coded for by the part of the gene holding the altered base group. In many cases where only one or several amino acids in a peptide chain of an enzyme molecule are changed, the enzyme keeps much of its normal function, since such is not a major change in the structure of the molecule. All this was only supposition on Crick's part, but it seemed very sound.

In the case of mutations that totally destroyed the workings of a gene, Crick assumed a base group was added or deleted near the middle or beginning of the DNA chain of the gene. The message would be read correctly up to the base group added or deleted. But from that point on, it would read incorrectly, and most of the resulting peptide chain would have the wrong amino acid units. The enzyme molecule, of which the chain formed a part, would not perform its normal function; the enzyme would not work. Crick could not tell whether such complete mutations were caused by an addition or deletion of a base group, although the possibility that they were caused by one or the other was strengthened through other findings with T4.

Part of this research had to do with cases where more than one such mutation were present in a gene of a virus. The DNA core of the virus had been mapped quite accurately, and Crick was able to artificially induce more than one nonfunctional mutation close together in a gene of its DNA core. He would grow a colony of T4 having one such mutation and another colony having another such mutation. He would then let both kinds of T4 infect bacterial cells at the same time, so that a few doubly mutant T4 having both mutations would arise. He then noted a peculiar fact: while some of the doubly mutant T4 came with the gene totally inactive as it had been in the singly mutant ones, others came with the gene almost as active as in the normal T4 virus. There was nothing really surprising about the cases in which the gene was inactive. One would expect that two mutations, both rendering a gene inactive when present alone, should make the same gene inactive when both are present. But often Crick found cases in which two such mutations made the gene almost as active as the normal unmutated gene; this was totally unexpected at the time. Often, two fatal mutations seemed to cancel one another when present together in a gene. It was also found that the two gene mutations had to be close together in the same gene if such was to be the case; otherwise they would render the gene inactive as each had done when present in the gene alone.

PLUS AND MINUS MUTATIONS

All these facts could be explained, thought Crick, if two types of mutation existed. One of these, a + mutation, would consist of the addition of a base unit to the DNA chain of the gene, while the other, a − mutation, would consist of the deletion of a base unit from the chain. If two + mutations or two − mutations occurred far apart in a gene, they destroyed its normal function, although if a + mutation and a − mutation occurred close together in the gene, the gene functioned almost normally, as if the two opposite kinds of mutation (one being the addition of a base unit and the other a deletion of a base unit) somehow cancelled each other's effects.

Crick found many pairs of mutations in the same gene of the T4 virus that, when close together, gave almost a normal virus. But often two mutations close together in a gene of the virus destroyed the function of the gene. These were surprising findings! Such patterns showed up only when the mutations were close together on the gene since, if they were very far apart, a combination of a + and a − mutation also made the gene totally inactive. Why? Through these observations, Crick found clues about how the genetic code works.

Before turning to that, let's see how Crick interpreted + and − mutations. The chief clue here was that the + and − mutations had to be close together for the gene to show normal activity. So he reasoned that if a − mutation represented the deletion of a base unit, while a + one represented an addition of a base group, his findings would have a simple and logical explanation. It goes like this.

Consider two + mutations close together on a gene. If, at each of these mutations, a base group has been added to the DNA chain, the genetic message would be read correctly until the first added base group is reached. At that point and beyond, the message would be read incorrectly, even if another added base group did not follow, although one actually does; and that only makes matters worse. The same holds true of two close − mutations or base deletions.

But what about two closely spaced mutations, one of which is a + mutation and the other a − mutation? To see the answer, Crick envisioned the machinery of the cell reading the message from one end of the DNA chain to the other. This implied that the order of the base groups determined the message. One thing seemed certain: a change in the order of the base groups changes the message the machinery of the cell reads, and leads to a gene that makes a different peptide chain than the normal gene makes. If we interpret a − mutation to be a base deletion, and a + mutation to be a base addition, as Crick did, we have an explanation of this finding. Let part of the base arrangement of a gene that holds closely spaced + mutations and be as shown below in its normal form without mutation. Now let the addition and deletion take place in the segment holding the six base groups between the bars. Remember, this polynucleotide chain is very long and can hold hundreds of base groups so that the segment between the

. . . . AGTCAC/TTGCAT/GAATCGATA . . . .

bars is a small part of the chain. If the base arrangement is as shown, the cell reads the message of this gene correctly, and the normal peptide chain will result.

Now consider the six base groups between the bars, or the sequence:

TTGCAT

Let the first base group at the left, a T group, be deleted, while another base group, an A group, is added between the last two groups—the A and T groups—at the right. Then the segment of the chain between the bars becomes

TGCAAT

which is different from that above. But this type of addition and deletion of base groups in a small part of a gene represents just the interesting situation Crick studied. So before the double mutation, the gene is as shown by the first point. After the double mutation it is like the second. Now with the first, the correct message is read by

1. . . . . .AGTCAC/TTGCAT/GAATCGATA . . . .

2. . . . . .AGTCAC/TGCAAT/GAATCGATA . . . .

the cell and the gene functions normally. But with (2) the message is read correctly up to the first bar and after the second bar, while, in between the bars, it is read incorrectly. However, the bases between the bars make up only a small portion of the gene. Thus most of the message is read correctly, and the peptide chain produced has only a small number of amino acid units different from the normal one: not a big difference in any case. So the gene under consideration could be said to work almost normally when a + mutation and a − mutation are close together in it. That was Crick's main reasoning, although the details of how the base groups code for amino acid units were unclear.

It was also simple to see that a single addition or deletion of a base group in a gene should render the gene inactive, if it is not near the end of the gene. Say the DNA chain of a gene is in part:

. . . . AGCTAA/GCTATAGGATAT . . . .

The bar marks the point where a base addition or deletion is to be made. Let the G group to the right of the bar be deleted; that is, say a − mutation takes place within the gene, so that the resulting DNA chain would then be:

. . . . AGCTAA/CTATAGGATAT . . . .

Also let the bar be at a point near the beginning or middle of the gene. Then, to the left of the bar, the message will be read correctly, since the base arrangement there is the same as in the normal gene with no mutations. But, to the right of the bar, the base arrangement is different from that in the normal gene. So beyond the bar, the cell will read the genetic message different than it would normally, leaving half or more of the peptide chain coded for by the gene with the wrong amino acid units, and the gene certainly would not function in its usual way.

Likewise imagine that an A group is added at the right of the bar to give the DNA chain:

. . . . AGCTAA/AGCTATAGGATAT . . . .

Again, the message to the left of the bar would be read correctly, while that to the right would be read incorrectly, and again, the gene would not function. So a + or − mutation alone would render the gene inactive.

Now say the G base group to the right of the bar in the normal gene is replaced by a T group that is, a mutation occurs that is a substitution of one base group by another. Then the DNA chain of the gene becomes:

. . . . AGTCAA/TGCTATAGGAATAT . . . .

Again, to the left of the bar, the base group arrangement is the same as in the normal gene. But what about the base group arrangement to the right of the bar? Remember the assumption that the machinery of the cell (whatever it eventually proved to be) reads the genetic message from one end of the DNA chain to the other; this had not been definitely proven yet, although Crick thought it to be a sound assertion. Then, to the right of the bar, the base arrangement would be the same as in the normal gene, except for the one base unit at the start of the sequence, the T group. It is therefore reasonable to assume that all the genetic message to the right of the bar would be read correctly, except for the small part of the message at the start of the sequence, with the result that probably only one amino acid in the peptide chain of the mutated gene would be different from the corresponding one in the normal peptide chain. Again, the mutated gene should function almost like the normal gene.

Most of the Crick's findings with T4 could be explained in these ways. But he hit upon another strange phenomenon; it concerned mutant T4 viruses that held three or more mutations close together in a given gene. He grew a colony of T4 having two + mutations close together on a gene, and another colony having a single + mutation in the same gene near the two mutations in the first colony, and then allowed both types of T4 to infect a culture of *E coli* cells. In this way he obtained T4 having three + mutations close together in the same gene through crossover. One property of these triply mutated viruses caught his attention: the triply mutated genes functioned almost normally, or were at least partially active. The same held true if the + mutations were replaced by − mutations. What might this mean? Two − or two + mutations close together in a gene destroyed the activity of the gene, but three + (or three −) ones close together made the gene substantially active. Very strange indeed! It would seem that any number of the same kind of mutation in a gene would destroy the function of the gene, though such was not so according to Crick's discoveries. Actually he was not experimenting with T4 through mere curiosity, but had definite ideas about how the base arrangements on the DNA chains of a gene held the genetic message that determined the particular peptide chain the cell makes from that gene, and was testing them in his experiments. From the standpoint of his theory, it was expected that three + (or three −) mutations close together would make a gene at least partially active. But to one with less experience in biochemical theory than Crick, one of the noted pioneers in the field, it seemed strange indeed.

THE GENETIC MESSAGE

This observation helped provide an answer to a question left unanswered earlier: how does the base arrangement along the DNA chains of a gene carry the genetic message for the making of the peptide chain governed by the gene? Clearly there must be some correspondence between the base arrangement of a gene's DNA and the amino acid arrangement of the peptide chain of that gene. What exactly is the nature of this correspondence? We have seen that a single base group cannot correspond to a given single amino acid. There are not enough base groups, and one may wonder at this stage: why not decipher the base arrangement of a given gene directly and compare this with the amino acid arrangement of the peptide chain the gene controls? That should quickly solve the problem.

There are some significant stumbling blocks to this direct approach, even today. First, as we have seen, genes are buried deep within the cell nucleus or in the core of a virus, and although chromosomes have been accurately mapped in some simple cases, they have not been studied in complete detail in larger organisms. For most large organisms only the location of certain genes on a given chromosome had been determined, and even the locations of many important genes in these organisms were not known with much precision. But secondly, even in Drosophila, one-cell organisms, and viruses, where complete chromosome mapping is possible, it was still not practical—it is still difficult today—to isolate a large number of a given kind of gene intact in the test tube and analyze it for its base groups and their arrangement. Chromatin of cell nuclei holds many different genes which cannot be separated from one another on any large scale in an intact form, although such would be required if one wanted to decipher the exact base arrangements of specific genes. Besides, even when measurable amounts of DNA from a given gene are available, the working out of its base group sequence is a challenging problem in chemical analysis and is very time-consuming, even with the aid of automatic, high speed devices in the modern biochemist's laboratory. It has been done in only a relatively few cases. The same comments apply to the analysis of a given protein. The exact amino acid arrangement has

has been figured out only for certain peptide chains of enzymes and structural proteins—for example, insulin, oxcitocen, and hemoglobin. A direct comparison of base sequences and amino acid sequences was out of the question in the late fifties and early sixties, and more indirect methods had to be employed to tackle the problem. That is where Crick's discovery that three + or three − mutations close together make the gene work (or partially work) was particularly enlightening.

The idea that one base unit stood for one amino acid unit was out. However, it seemed clear that if a specific base group arrangement determines the instructions for the creation of the peptide chain controlled by a gene, the cell must somehow start at one end of the DNA chain and read the message by noting a certain number of bases in succession. First, it might take note of the first two base groups at one end of the chain, then the next two, then the following two, and so on, if it reads two groups at a time. But is the genetic message read two groups at a time? What is the actual number read each time? For now, assume that two base groups at a time are read. In other words, say that each amino acid is represented by a combination of two base units next to one another on the DNA chain. How many different ordered pairs of two base units can be constructed out of the four different base units? This question had to be answered to decide if there are enough such pairs to accommodate the 20 amino acids.

The general form of a combination of two base groups next to one another on a DNA chain is BB′, where B can be any of the four base groups A, T, G or C, so that there are four possibilities for B. Also B′ can be any one of the four bases, again making four possibilities for B′. This makes 4×4, or 16, different ordered pairs of the form BB′, and therefore, there are 16 pairs or combinations of two adjacent base groups possible at each site on a DNA chain which, when written out exactly, are:

AA	TA	GA	CA
AT	TT	GT	CT
AG	TG	GG	CG
AC	TC	GC	CC

Note that AT, for example, is different from TA, since the order of the base units is important in the DNA code. So if the machinery of the cell reads the bases two at a time down the DNA chain, there will be 16 two-base combinations it could read in all. Let each such two-base, ordered combination stand for one of the amino acids. Then there are only 16 combinations, but 20 different amino acids. Again, this scheme solves nothing. There are still too many amino acids to be accommodated by the two-base DNA code.

What happens if we assume, as did Crick and his colleagues, that the genetic message is read three base groups at a time? The general three-base combination has the form BB′B′′. Again B can be any one of the four base groups as can B′ and B′′. Thus there are $4 \times 4 \times 4$, or 64 different three-base combinations. Now assume that each such combination stands for one amino acid. But there are 64 three-base combinations and only 20 amino acids. How could this situation be interpreted? If one assumes one- or two-base combinations, there are not enough combinations for the 20 amino acids, while if one assumes three-base combinations there are too many. However, imagining the message to be read four, five, or more base groups at a time was no way out of the difficulty, since then there would be far too many combinations—many more than 64. Such base group combinations became known as *codons*, and each codon would code for an amino acid.

Was the puzzle really as strange as it seemed? Clearly, a real problem existed if there was not enough codons (as in the case of one- and two-base codons) to accommodate the 20 amino acids. The message is read passively. Each codon would have to stand for one amino acid only. If there were only four or 16 codons, no codon would exist for some amino acids. Yet it was known that all the amino acids occur in peptide chains, so each amino acid had to have a codon. What if each of the 64 three-base codons did in fact stand for a single amino acid? Would that really cause a problem? It would mean that more than one of the 64 codons could stand for the same amino acid, because there are only 20 amino acids and 64 codons, though all the machinery of the cell requires is that each codon stand for only one amino acid and no more; so all the 20 amino acids could thus be represented in the system of 64 three-base codons. But, then, what would restrict us to three-base codons? Would not codons of more than three bases do just as well? The answer is yes. And theory alone could not help here. One could suppose, however, as did Crick, that the simplest scheme is the best in science, and go on to assume that each of the 20 amino acids is represented by a three-base codon. That is a good starting place. But simplicity in itself is no basis for truth in science, and a simple explanation (like all) must be backed by the known facts revealed through experiment. Sometimes the solution proves wrong in light of experimental fact. The only way to settle the matter was to turn to experiment.

That is just where Crick's finding that three + or − mutations close together imparted activity to a T4 gene was of paramount importance. It implied that the cell read the genetic message three bases at a time. Consider the portion between the bars of the long DNA chain shown below made of 12 bases. Let this chain be the normal form of the gene. Now assume that the cell reads the message three bases at a time

. . . . AGTAGCTCG/GTACTTAAGTGA/ATGCGCCTA. . . . .

from left to right along the chain. The reading would start at the triplet codon AGT, which would represent a certain amino acid unit to be incorporated into the peptide chain of the gene, and the cell would somehow note this and proceed to the next triplet AGC. That triplet codon would stand for another amino acid unit to be incorporated into the peptide chain next to that represented by AGT. Then the reading would go on to the triplet TCG for the amino acid to be put into the peptide chain after those of the previous codons AGT and AGC. It would then go on to GTA, and so on, noting somehow that the amino acids are to be inserted into the peptide chain in the same order as their codons occur in the DNA chain of the gene. This all assumes that each triplet stands for only one amino acid. How would a mutation affect a gene in this scheme?

Consider the 12 bases between the bars. Imagine that the second, fourth, and eighth base groups between the bars—or T, C, and A—are deleted to give the resulting DNA chain:

. . . . AGTAGCTCG/GATTAGTGA/ATGCGCCTA . . . .

These deletions are the same as three − mutations close together in the gene. So the message of this triply mutated gene will be read correctly up to the first bar, although incorrectly between the bars—a total of three amino acids in the triplet code—while the correct reading will resume after the second bar, because there are still a whole number of codons— three to be exact—between the bars which are read one at a time, so that the first triplet ATG after the second bar will be read as in the normal gene. This means that all of the amino acids of the peptide chain that this gene governs should be correct, except for those corresponding to the codons between the bars, in the triply mutated gene of this example. The rest of the peptide chain should be normal. That is not a big difference. The mutated gene should behave almost normally. In the same way we should find that three + mutations in the bars would have a similar effect on the gene.

Now what should Crick have found with three + or − mutations close together in a gene if the genetic message was read, say, four base groups at a time? Look at the above triply mutated gene again:

. . . . ACTAGTAGCTCG/GATTAGTGA/ATGCGCCTA . . . .

Here three more bases have been shown on the left to simplify the situation. In the normal gene the message would be read four base units at a time in the order ACTA, GTAG, CTCG, GTAC, TTAA, GTGA, after which the second bar is reached and the reading would go on normally. But how about the gene with three − mutations between the bars? The first codons ACTA, GTAG, and CTCG are read correctly; but, then, the reading goes incorrectly with the codons GATT, AGTG, AATG, and so on, through the rest of the chain or gene. Thus, if the message is read four bases at a time, a combination of three + or − mutations close together in the same gene should render the gene inactive. But such is not the case as Crick had shown. His work therefore offered the strongest evidence that the genetic message is read three bases at a time.

Two + or − mutations close together in a gene renders the gene inactive. Go back to the normal DNA chain of the last example:

. . . . AGTAGCTCG/GTACTTAAGTGA/ATG . . . .

Consider two − mutations in which the second and tenth base groups between the bars—both T groups—are deleted. Then the chain becomes:

.... AGTAGCTCG/GACTTAAGGA/ATG ....

The order in which the triplets are read in the normal gene from left to right is:

.... AGT, AGC, TCG/GTA, CTT, AAG, TGA/ATG, ...

while in the doubly mutated one it is

.... AGT, AGC, TCG/GAC, TTA, AGG/AAT,.............

which is the normal reading up to the first bar but not after. In this doubly mutated gene, the reading does not come back in phase with the normal one after the second bar like it does in the triply mutated gene.

If the genetic message is read in base triplets, or in any whole-number multiple of three bases like six, nine, twelve, . . ., base additions or deletions close together in the gene that are triple mutations should bring the reading back in phase with the normal one after the segment of the DNA chain containing them is read. Any other number of such mutations like one, two, four, five, seven, etc., should put the reading out of phase with the normal reading after they are read. This means that the message read will be incorrect for much of the chain if the gene has one, two, four, . . ., + or − mutations close together near the start or middle of the gene. The gene will not function in that case. Experiment confirmed these predictions. There is little doubt among biochemists that the genetic code comes in "words" of three-base groups each.

It can also be seen why a + or − mutation (or a base addition or deletion) close together in a gene makes the gene active or partially active. If the number of bases between the bars in our example is the same before and after mutation—as it is when a + and − mutation is present there—the reading of the message will come back in phase with the normal reading after the bars, and will be mostly correct. The gene will therefore function almost normally. But if the number of base groups between the bars is changed by one, two, four, . . ., and so on, the reading cannot get back in phase with the correct one after the second bar. Then most of the message will be wrong. The peptide chain controlled by the gene will contain the wrong amino acids for the most part. The gene will not function.

CODONS AND AMINO ACIDS

All evidence at present implies that the chemical machinery of the cell reads the genetic message three bases at a time. Each of the 64 triplet codons may be called a word in the language of the DNA code. The language of protein, on the other hand, has 20 words, the amino acid units of the peptide chain. Crick's work mainly showed that each three-base word of the DNA language stood for one amino acid unit. But which amino acid did each codon of the DNA code stand for? All that one could say at this point was that each codon stood for an amino acid. But what amino acid in each case? This question posed the problem of translating the DNA language into protein language, and its solution will be dealt with in the rest of our story of the chemical makeup of the gene.

There was an immediate problem that occupied the minds of many biochemists in the late fifties and early sixties. It had to do with another peculiarity of how the cell reads the DNA language of the chromosomes. Consider a DNA chain like that below. Does the cell read the codons or base triplets one after another

. . . . AGTCATGCGATTGGTTCACGTA . . . .

in the fashion AGT, CAT, GCG, ATT, and so on? Or does it do so in another way such as AGT, GTC, TCA, CAT, ATG, . . . , starting the reading over at each base unit on the chain? The first way of reading the genetic message was known as *nonoverlapping*, while the second was known as *overlapping*, since each pair of adjacent triplets read have two base units in common, and most base units of the chain are read more than once. It was not clear at first in which of the two ways the cell reads the genetic code. So the question becomes: Was the code overlapping or nonoverlapping? It has been assumed that the reading is nonoverlapping for simplicity. However, the explanations of Crick's findings would be much the same in an overlapping code. Actually the question was not hard to settle. In fact, it was answered at about the same time that the triplet nature of the code was being tested.

If the code is overlapping, certain facts can be predicted that can be checked by experiment. Notice, for example, that each of the triplet codons below of

(1) AGT
(2) GTC
(3) TCA
(4) CAT
(5) ATG

our example have two base units in common. The last part of (1), the GT part, is the beginning part of (2). Also the last part of (2), the TC part, is the beginning part of (3), and so forth. In other words, each codon read in the overlapping code should in part determine the codon read next. It therefore seemed logical that certain amino acids should often follow others in natural peptide chains. Such was not the case. Which amino acid followed others in actual peptide chains was a random matter on a large scale. Amino acid groups showed no preference to follow others in real peptide chains.

More important was another observation about mutations. The smallest effect of a mutation is on a single base unit. When such a small mutation takes place, only one amino acid in the peptide chain governed by the gene should be different from those of the normal peptide chain in a nonoverlapping code. But in an overlapping one, at least two or three consecutive amino acids in the peptide chain should be changed by such a mutation. But in many mutations only one amino acid of a peptide chain was found to be altered. This provided good evidence that the code is read in a nonoverlapping manner.

After Crick's work with T4, the idea that each amino acid of the peptide chain is represented by a three-base codon on the DNA chain won acceptance. Also the DNA code came to be seen as read in a nonoverlapping way. Today both these assertions are regarded as facts by biochemists and geneticists. But one more idea should be mentioned: some biochemists at first thought that only 20 of the 64 codons represented amino acids, while the rest would stand for nothing and would be "nonsense" triplets. But there is one problem with that idea: how would the cell interpret a nonsense triplet? Clearly many of the triplets spanning the length of the gene—44 of the 64 possible—would be nonsense. The most reasonable assumption was that the reading of the message would stop the first time a nonsense triplet is encountered. But then most, if not all, genes would not function. All evidence spoke against this. So biochemists came to assume that a vast majority of the 64 codons stood for amino acids, although there are a few nonsense codons. The cell can live with a small number of them. In fact, they are essential to the DNA code.

The determination of how peptide-chain information is coded for in the DNA molecule's genes, and how the genetic message given there is read, was only part of Crick's research and theorizing; it was only a beginning. Much work remained to be done on the problem of how protein and DNA language are related. What codons coded for a particular amino acid? How could that be figured out? These questions will be addressed later. But now let's review how far we have come in understanding what Mendel's genes really are.

CONCLUSIONS

When Mendel postulated the gene to explain his experimental findings with pea plants, he thought of it as some sort of physical factor in the plant's cells. But he knew nothing of cell organelles, chromosomes, or DNA. The only assertion he felt sure of was that the gene had to be a physical body of some kind. That was as far as anyone's understanding of the gene could go for the following 30 years, since the microscope was not sufficiently developed to permit accurate observations of cells and their component parts at Mendel's time or in the years before his historic work became known. So the gene idea did not take hold in the scientific community right away.

That changed, however, near the turn of the century when sections of Mendel's papers were uncovered by De Vries, Corens, and Seysenegg. But where in the cell were genes located?

At the turn of the century better microscopes had come into being, and many cell organelles could be seen more clearly, and followed through the process of cell division taking place in growing tissues of organisms arising from their embryos—the first cells of organisms formed by the union of the egg and sperm. One of these organelles, the chromosome, was particularly interesting. It behaved like Mendel's genes would be expected to behave in many ways. These observations gave rise to the short-lived idea that the chromosome might be the gene. It has been shown, however, that some simple facts dispelled that notion. Yet it seemed certain that genes had to be associated with chromosomes in some way, since, after all, chromosomes were small organelles in cell nuclei that behaved like genes in so many ways. Thus, the idea arose that the chromosome held many different genes.

But the belief that chromosomes held genes was not enough to explain many experimental findings, though it presented a clearer picture of the particulate or physical nature of the gene. The probing went on. The chromosome was finally mapped in simpler organisms and the gene came to be seen as a small segment of the chromosome. These developments marked the first era of modern genetics. The gene had gone from just being a physical factor to being a definite part of a visible cell organelle, the chromosome, having a behavior that could clearly be seen during cell division. This period lasted until around 1950. It was noted for far more than chromosome mapping, however. Biochemists were working out the basic chemistry of chromosomes at the same time. Also the experiments of Griffith and others, dealing with the transformation of bacteria, shed much light on this chemical problem. Because chromosomes held genes, it was clear that the genetic substance transferred in these experiments had to reside in them, while biochemical studies had shown that chromosomes were made of two kinds of substances: DNA and protein. The question then arose as to which of these substances was the genetic material. Experiments such as those of Avery in the mid 1940s went a long way in showing that DNA is that material.

Then, in the early 1950s, came the Watson-Crick model of DNA structure, marking it apparent that the secret of the gene resided in the base arrangements of the DNA molecule. This revelation marked the beginning of what may be called the second era of modern genetics. In this period the gene went from being seen as a small segment of the chromosome to being seen as a certain arrangement of the bases on the DNA chain in the chromosomes. This interpretation was given much experimental support by the work of Francis Crick in the late 1950s.

However, what precisely is the gene in this model? The experiments of Beadle and Tatum had taken a big step to answering this question and helped show that the entity Mendel thought of as a single gene controlled the making of a certain peptide chain needed by the cell. Thus, what Mendel called a gene is really a sequence of bases along a DNA chain that governs the making of some peptide chain. The peptide chains in the average cell come in many different sizes. So the DNA chains that make them must also. But the average sequence of bases representing a Mendelian gene is around 900 base units long. The machinery of the cell begins at one end of the sequence of base units and reads the base triplets (or codons) one after the other, somehow noting which amino acid each codon stands for, and then carries instructions for the assembly of their amino acids into peptide chains at the ribosomes of the cytoplasm in the same order that the codons are read along the DNA chain. Mutations can be seen as additions, deletions, or substitutions of base groups in the DNA chain. The study of mutations greatly helped biochemists arrive at the modern view of the genetic code. Solving the problem would have been much more difficult without gene mutations and the ability to bring them about in the laboratory.

Many questions remained. For example, peptide chains are assembled in the cytoplasm at the ribosomes. How is the genetic message transferred from the DNA of the nucleus to the ribosomes? All that Crick's work with T4 implied was that each base triplet in the chromosome corresponds to some amino acid unit. That research, by itself, said nothing about the means by which the message is carried to the cytoplasm. But there were speculations. It is a long story that will be taken up at least in part in the rest of this book.

The cell somehow reads the DNA message of the gene. But, remember, Watson and Crick showed that the DNA molecule has two chains or strands. Which of the two strands does the machinery of the cell read? Does it make a difference which one it reads? Biochemists managed to answer such questions too.

Also recall it has been shown that chromatin contains a lot of protein—but a particular variety known as *histones*. What is the function of histones in chromatin? Biochemists have made significant headway in finding out.

An additional question is: How does the chemical machinery of the cell know where one gene ends and another begins on the DNA chains of the chromosomes? What signals the end of one gene and the start of another? There must be something that does this, or separate genes could not exist. But what? This question, too, will be dealt with in what follows.

These are some of the questions that had to be answered in the mid 1950s. Although the first step to discovering how the genetic message gets from the nucleus to the ribosomes had to do with the second kind of nucleic acid, RNA, only DNA has been covered most extensively so far. That is because DNA is the chief genetic material. But RNA has a very important and indispensable function in the heredity of the organism. When chromatin of cell nuclei was analyzed chemically, RNA was found there too. You have probably been wondering what specific purpose RNA serves. The role of DNA in cells was more obvious because it formed the chief part of the chromosomes visible under the microscope that acted like Mendel's genes, and had to be involved in the process of heredity directly, while the exact location of RNA was not clear at first. Biochemists had to probe harder to find its locations in the cell and the role it played in carrying the genetic message from the nucleus to the cytoplasm of the cell.

However, our present knowledge of the role of RNA only came after geneticists, biochemists, and cytologists, scientists who study individual cells and their structure, used certain methods to probe cell organelles which have not yet been explained. The study of cells and their chemistry is a complex affair. The question of the purpose of RNA in the cell makes it necessary to look at these methods more closely. The closer one can get to the daily activities of scientists making historic discoveries, the easier it is to understand how discoveries are made. At the same time, valuable insight into their importance in the framework of a given science is gained.

We are still living in the second era of modern genetics. Who knows where it will eventually take us?

Part 2

The Methods of Biochemistry and Genetics

Chapter **6**

Nineteenth century methods

Much has been said about chromosomes and other cell organelles. But how were life scientists able to see the finer details of such small structures and work out their chemical makeup? You might think the answer is simple: they simply looked at living tissues under high powered microscopes and could see cells and their parts easily. In an overall sense, that is true. However, such a procedure was not as simple in practice. There was one problem: cells and their organelles as seen under a light microscope are almost as transparent as window glass. Many details of their structure were not clear under the first microscopes of the seventeenth and eighteenth centuries. That was a serious problem for investigators of Mendel's day and before.

In the second part of our account, let's look at some of the methods of cytology, biochemistry, and genetics of the past, and work up to modern methods and what they revealed about the genetic code and its finer details. The approach of this part of the book will illuminate additional methods of modern biochemistry and genetics and the discoveries they made possible, and will also be more chronological in its approach to give the reader a clearer picture of the study of heredity and how it developed up to the present time.

In Part 1, the main findings, developments, and important methods of biochemistry and genetics were given. Thus the reader should have the key concepts of the field clearly in mind at this point. Part 2 explores the *how* and *why* of modern biochemical and genetic knowledge. The questions posed at the end of Part 1 are resolved here. The promise modern genetics holds for the future is also explored.

THE MICROSCOPE BEFORE THE TWENTIETH CENTURY

Probably no instrument has influenced the development of life science more than the microscope. This device of scientific investigation has a long history. Magnifying glasses were known in ancient times. Yet real progress in making better magnifying devices did not begin until the scientific revolution of the sixteenth century, initiated by the Italian investigator Galileo Galilei and the great English scientist Isaac Newton. Galileo's first telescope and magnifying lenses were actually primitive compared to those of today's technology. But in their day they marked a giant step forward and inspired the biological advances of the seventeenth century. Some of the first real microscopic examinations of living organisms were made in the seventeenth century.

Before looking at these, let's briefly review some of the main reasons scientists turned the microscope to living organisms. From the beginning of written history, they could see some basic differences between dead matter making up stones, water and sand, and living matter making up people, animals and plants. The ancient Greeks wondered about the basic nature of dead matter. Some Greek philosophers thought such matter was continuous. Others, like Democritus, believed that inanimate matter was not continuous, but consisted of very small indivisible particles, or atoms, that seemed to account for much of the observable behavior of dead matter. Many atomists, as the believers in this theory came to be called, also thought living matter consisted of atoms. But prior to the eighteenth century, it was a common belief that living matter was in some way inherently different than dead matter.

Although atomic theory did not prosper greatly in western thought for more than two thousand years after Democritus, there is evidence that the founders of the modern scientific method in the sixteenth and seventeenth centuries had some leaning toward the theory. Some of Newton's writings indicate he believed in matter composed of atoms.

So, just as speculations about the nature of inanimate matter had flourished in all ages, scientists of the seventeenth century began to probe the nature and constitution of living matter. Here the development of the first microscopes came in handy.

To go into the technical details of even the simplest design of these first microscopes would be impossible here. This is a branch of technology to which an entire book in itself could only begin to do justice. But any microscope has two properties that are important to the work of the biologist. One of these is *magnification*. The magnification of a microscope is merely how many times the instrument enlarges an object it is used to examine. If viewing a small object with the instrument makes the object appear 20 times as large as it actually is, we say the microscope has a magnification of 20. Another property of a microscope is its *resolving power*. We will say more about this later. Magnification was most important with the first microscopes of the seventeenth century.

In that century biologists began to probe the finer details of organisms with the microscopes. The first microscopes of the previous century only had a magnification of 10. Yet they could make organisms barely visible to the naked eye show many details that were not even imagined before their use. One investigator, Francesco Stelluti, made intensive studies of the bumble bee, beginning in 1625, using magnifying lenses having magnifications of only 5 to 10. But, surprisingly, he could see and study many parts of the bee not visible without the lenses, even at such low magnification. Stelluti's investigations helped open up a whole new area of research. With such early innovations, the microscope became the chief tool of biological research.

Another noted observer and expert at making versatile, small microscopes to study living organisms in the seventeenth century, Anton van Leeuwenhoek of Holland greatly extended the studies of Stelluti and others. Leeuwenhoek was very gifted at the art of grinding small lenses and making little microscopes from them. Their magnification is reported to have been from 50 to 200. But the discoveries he made with his small microscopes were truly revolutionary. Two of them were particularly significant. He looked at rain water that had been standing for some time, and at water from ponds of various sorts. What he uncovered under his small microscopes about such water was unbelievable. The water was teeming with living organisms too small to see with the naked eye. Prior to Leeuwenhoek, no one dreamed these small organisms existed. However, a wide variety of them were uncovered under the microscope. Leeuwenhoek also made microscopic studies of the structure of tadpoles and found that his small microscopes not only made arteries and veins visible in these creatures, but also revealed that capillaries—small vein-like channels—connected arteries and veins, at least in the body of the tadpole. By means of these, blood circulates between the arteries and veins. This noted investigator made many other discoveries with his small microscopes. But these would take up a chapter in themselves.

Another seventeenth-century investigator who made history in microscopic studies was the Italian Marcello Malpighi. He made revolutionary observations of the lungs of the frog. Under the microscope he observed that the lung of the frog consists of little air-filled sacs.

These are just a small number of the many kinds of microscopic studies of insects and other organisms made in the seventeeth century. Because of them, that century began to see the rise of real scientific biology. But it was only a start. No great strides were made in microscopic investigations of living organisms in the eighteenth century. Instead, biologists of that century turned their attention to the classification and organization of living organisms. There were two main reasons for this halt in the progress of microscopic technology in that century. One had to do with a phenomenon known as *spherical aberration*: the light rays from the object viewed under the microscope did not focus properly as the magnification of a given kind of microscope was improved. As a result, the object viewed seemed blurred and distorted under the microscope. This became a problem as seventeenth-century microscopists tried to improve the magnification of their microscopes.

Another was *chromatic aberration*. It can be understood by looking at a simple experiment Isaac Newton performed with the aid of a prism. A prism, like a lens, is a piece of transparent glass, but is triangular in shape. Newton let a beam of white sunlight fall on one side of the rectangular faces of a prism. On the other side of the prism was a white screen. After the light beam went through the prism and came out of the opposite rectangular face of the prism, it fell on the white screen. But what Newton found on the screen was not a white image of the beam that went into the prism. Instead he found a series of continuous bands of six different colors—red, orange, yellow, green, blue, and violet—on the screen. White light had been shown to be a mixture or blending of the light of six *primary colors*. Light rays of the six different colors are bent to different degrees as they enter the prism, and thus emerge from it separated from one another, while the same kind of thing happened to light going through microscopes as attempts were made to increase their magnification in the seventeenth century. Light rays of different colors were bent different amounts as they passed through the lenses of the microscope. As a result, when investigators looked at small objects through their microscopes, the objects appeared surrounded by rings of different colors which only served to distort the true nature of the objects being examined; this phenomenon is known as chromatic aberration. It was a stumbling block to microscopic technology in the seventeenth century. That is one reason why microscopic investigations did not make further headway during the eighteenth century. This was a problem with the technology of the times and proved solvable later.

By the beginning of the nineteenth century, the problem of both spherical and chromatic aberrations had been solved and light microscopes of much larger magnification than those of the seventeenth century were available by 1820. These developments enabled biologists to probe further into the structure and nature of living organisms. But, in the early eighteenth century, many microscopic studies of plants had been made. Recall that back in the sixteenth century Robert Hook looked at cork under a microscope and found that it consisted of small compartments or chambers he called cells. Now cork is the product of a plant. The chambers Hook observed were the remnants of the cell walls that enclose all plant cells. Each plant cell is enclosed by a wall; this was one fact that eighteenth century observers were able to see as they examined many different plants at different stages of their life cycles. The cell wall served to make plant cells more easily observed under the microscope. It was harder to see animal cells under the microscopes of the times. Animal cells do not have cell walls in most cases, and are often smaller than plant cells, and, for these reasons, the belief that all living matter consists of cells had not yet taken hold in the early 1800s.

By 1800, the French physician Francois Xavier Bichat had shown that the organisms of living creatures are composed of various tissues, and microscopic studies bore this out. Yet many investigators at the time thought they had found the most basic components of living matter in tissues, and did not come to the belief that all living tissues are composed of much smaller, microscopic units, or cells, as we now call them. But the observation of cells in many plant tissues started to give more credibility to the idea that cells might be the universal components of all living tissues, both plant and animal. Yet with the state of microscopic technology at the time, it was difficult in many cases to tell whether or not actual cells were being observed in many tissues, or whether the appearances were just distortions, or artifacts, caused by imperfect microscopic investigations.

The picture started to look brighter for the cell hypothesis by the 1830s. By 1830 cells were clearly visible in plant tissues seen under the newly improved light microscopes; the cell walls surrounding them made this possible, although the cells, for the most part, appeared transparent under the microscope and thus did not reveal much internal structure. But, in 1831, the Scottish botanist, Robert Brown, using a microscope with a magnification of 300, could make out a more compact region in the vicinity of the center of many plant cells that he called the *nucleus*. This proved to be an important discovery. In 1838 the German scientist Mathias Jakob Schleiden put forth the idea that the nucleus must be an important structure in the plant cell, since it was observed in many of them. Probably taking into account the drawbacks of the microscopic techniques of his day, he postulated that the nucleus was a part of all plant cells. But what about animal cells? In 1838, microscopes with a magnification of 450 were in common use. These were good enough to make cells visible in many animal tissues, although such cells were harder to see, because they were not enclosed by cell walls and were also usually transparent. The nucleus had also been seen in many animal cells. These observations got another German scientist, Theodor Schwann, to postulate the presence of the nucleus in both plant and animal cells, and he came to see its presence as the clearest proof that both plant and animal tissues consist of cells. Today, Schleiden and Schwann are considered the chief inventors of the cell theory.

Another matter of great interest to biology was investigated near the middle of the nineteenth century: that of the development of the organism from its first cell, or *embryo*, that gave it birth. With better microscopes, scientists were able to observe the development of the tissues of various plants from that first cell. They had done the same with animal tissues by the middle of the nineteenth century. The process of *fertilization*, the union of egg and sperm, had also been observed in certain cases. The egg cell was seen to be much larger than the sperm cell, although most of the egg was shown to be made of nutritive material with a very small cytoplasm and nucleus. The sperm cell was seen to be made of a head which contained its genetic material, and a tail-like portion that enabled it to move toward the egg cell and unite with it. The process of meiosis, or germ cell formation, was also being scrutinized under the microscope in some special cases. All these breakthroughs through the use of the microscope led to the full acceptance of the cell theory of living matter by the second half of the nineteenth century.

Prior to the middle of that century, there was still one problem when it came to studying the organelles of animal cells. That is, they were transparent and the organelles in them could not be seen or distinguished from each other. Also animal cells did not have cell walls to mark them off from one another.

Both problems were solved when microscopic studies were aided by the discovery of new dyeing agents to color cell organelles so that they stand out under the microscope against their transparent background. By 1870, dyes that colored each kind of organelle differently were in common use. Each such dye imparted to given kind of organelle a different color. Then different organelles could be distinguished by their different colors when tissues were treated with different dyes. In this way chromosomes were discovered and followed during cell division in the latter part of the nineteenth century. Also, approximate chromosome counts were made in various organisms, and it was becoming clear that a given kind of organism had a fixed number of chromosomes in its body cells, while its germ cells were observed to have half that number of chromosomes. By 1900 the existence of the chromosome was accepted as a reality.

Cell theory had brought much order to biology. But the theory owed its verification to the newly improved microscopes of the nineteenth century, while the study of chromosomes shed much light on cell heredity.

However, even in the eighteenth and nineteenth centuries the study of living matter progressed along another line that eventually led to our modern understanding of the gene and how it is constituted: that of chemistry.

BIOCHEMISTRY BEFORE THE TWENTIETH CENTURY

In the latter half of the eighteenth century, chemistry was making strides at becoming an exact science. At that time the great French chemist Antoine Laurent Lavoisier had established that combustion consists of the union of the burning substance with oxygen of the air. This discovery discredited the then prevailing phlogiston theory of combustion, which held that metals and other substances that burn, or combust, gave off a rather mysterious fluid known as phlogiston as they burned. If that were so, the remains left after the substance burned should have weighed less than the original substance. However, Lavoisier and others found that the material left after some metals burned weighed more than the original amount of metal burned. He proved experimentally that when a metal undergoes combustion, it unites with oxygen of the air to form another substance made of that particular metal and oxygen. But he also showed something much more important. The sum of the weights of the metal burned and that of the oxygen consumed in the burning equals the weight of the substance left after burning—the metallic oxide in this case—showing that no matter was created or destroyed in the process of burning. Lavoisier made other studies which showed that no matter is created or destroyed in any chemical change. This law of conservation of matter made chemistry an exact science.

In the late eighteenth century, chemists were turning their attention to the chemistry of living matter. There seemed to be some striking differences between the chemical behavior of living matter and that of dead matter. To make this clear, consider water and salt. These are two substances that are typical of the mineral realm, distinct from the plant and animal sphere. Both substances are very stable under heat and pressure. Water can be heated until it vaporizes. Then the steam can be cooled until it condenses to give water again. Water will not undergo the chemical change of decomposing into its constituent elements, hydrogen and oxygen, unless it is subjected to very high temperatures and pressures in the gaseous state, or unless an electric current is passed through it. We can heat table salt intensely over a hot burner of an electric range, or in the hot flame of a bunsen burner, but it will not even melt. It will melt only if subjected to high pressures and temperatures at the same time. But, even then, it will not separate into its elements, sodium and chlorine, unless an electric current is passed through it. Water and salt are very stable under heat and pressure. The same applies to most substances like sand, limestone, iron ore, carbon dioxide, and others obtained from the inanimate world of rocks, minerals, and air.

Contrast this behavior with that of two other substances like sugar and wood, both of which are obtained from living organisms. Wood will readily undergo the chemical change of burning. The same holds true for table sugar. When sugar is heated mildly it begins to turn into a brown liquid mass that finally turns black, because of its decomposing into carbon and water on heating. The same kind of observations hold for most of the substances in milk and butter, both of which are from living organisms.

Chemists of the late eighteenth century took note of this fragile nature of the substances making up living matter and compared it to the behavior of substances making up the inanimate world. One group of substances in living matter were particularly interesting in this regard. In addition to being very unstable under heat, they had another strange property. It is illustrated every time one fries an egg. Egg white before the egg is cooked is a thick gelatin-like liquid. But when mild heat is supplied, it hardens into a white soft solid, while the same is true for the yolk of the egg which is originally a viscous yellow liquid. On heating it hardens into a yellow soft solid. Most other substances, both from dead and living matter, melt or liquify on heating, while egg white and many other materials eighteenth-century chemists found in living organisms solidified on heating. These kind of substances seemed to make up most of the matter in the living organism. In 1777 the French chemist Pierre Joseph Macquer made a thorough study of this peculiar behavior. He thought of such substances as albuminous, after *albumin*, the ancient name for egg white, and this wide variety of substances in living matter became known as *proteins*. The name was derived from a word of ancient Greek origin that meant "having first importance," since it soon became apparent that protein substances were the most numerous in organisms and apparently seemed to serve important purposes in them.

At the beginning of the nineteenth century, various attempts were made to analyze protein substances. They were found to be much more complicated than carbohydrates and fats. Chemical techniques had developed to the point where the chemical elements in carbohydrates, fats, and proteins could be determined. Both carbohydrates and fats were shown to be composed of carbon, hydrogen, and oxygen, while by 1839 it was known that proteins contained nitrogen in addition to these three elements. By 1849 the work of the great German chemist Justus von Liebig showed that proteins contained both nitrogen and sulfur in addition to carbon, hydrogen, and oxygen, and therefore began to come to the conclusion that they were more important to the life process than the fats and carbohydrates, since they contained the most chemical elements of all substances in the organism.

In 1819 important aspects of the composition of plant starches and celluloses were being uncovered. Celluloses, like starches, are carbohydrates. But they are harder and sturdier than starches, so that they make up the stronger parts of the plant, like the wood of trees, that give the plant support. In 1819, the French chemist Henri Braconnot heated various plant starches and celluloses in acid solutions and found that such solutions dissolved them. What did this mean? When he separated the components of the resulting mixture, and the initial water and acid in the solution were accounted for, the remaining portion proved to be the simple sugar glucose. It therefore was concluded that the acid solution had separated the starches and celluloses into glucose. The only logical conclusion was that these carbohydrates were made of complexes of many glucose molecules. This was a needed break. It started to become apparent by 1820 that the molecules of substances making up living matter were more complex than the simple ones making up inorganic, or nonliving, matter. Braconnot's findings would later give chemists an ingenious clue as to how to work out their structure. But at the time chemists had no clear idea of how complex these organic molecules really were, so that Braconnot's work was only exploratory, though it did tend to show that all the various plant starches and celluloses had molecules composed of many glucose units. He did not stop there, however.

He next tried to find out what effects heated acid solutions would have on the proteins, or albuminous substances, of organic matter. Gelatin was one such substance known in his day. When he heated gelatin in an acid solution, he managed to collect crystals of a white, sweet tasting substance from the resulting mixture. Because of its sweet taste, he thought the substance might be a sugar. So he probed the composition of the substance further, and found that a simple nitrogen-containing substance, ammonia, could be obtained from it. Thus, the sweet substance had to contain nitrogen and thus could not be a sugar. The substance came to be called glycine and was one of the first amino acids discovered.

Next Braconnot heated muscle tissue in acid solution and obtained another substance made of white crystals. This event marked the discovery of the amino acid leucine.

Additional amino acids were separated from the albuminous substances, or proteins, as the nineteenth century progressed, and by the latter half of the century it became clear that the proteins were far more complex in composition than the fats or carbohydrates. The carbohydrates had been shown to yield only glucose, and some other simple sugars, on acid treatment, while the proteins gave many different simpler substances. The last of the 20 amino acids was not uncovered until 1935. However, by the last decade of the nineteenth century, chemists had made headway in working out the molecular and structural formulas of simple inorganic substances and the simpler organic ones like the simple sugars and amino acids. The structural formula of glucose was worked out by 1886 by the German chemist Heinrich Kiliani.

By the last decades of the nineteenth century it was obvious that the large molecules of fats, carbohydrates, and proteins are much more complicated in structure than those of inorganic matter. By the ordinary chemical methods of the time, it was easy to work out the molecular formulas of simple inorganic substances like water, ammonia, carbon dioxide, carbon monoxide, and so on. Although it was more difficult, the structural formulas of simple organic ones like glucose, fructose, and galactose, as well as those of various amino acids, were also figured out by these techniques. But for molecules much larger than these, the task was impossible. This was true especially for proteins, the most complex substances of organic matter. It was in the late nineteenth century that Braconnot's findings came in handy. Chemists had come to see what it really meant by that time. They realized that they did not have to tackle the problem of determining the structural formulas of these large molecules directly. Instead, the large molecules of carbohydrates like starches and celluloses could be seen as made up of chains of simpler molecules, the building block being glucose. The structural formula of glucose had been deciphered. Chemists then could understand how glucose molecules could join up with each other to form starch and cellulose molecules.

There was still a problem when it came to the proteins. While it had been shown that starch, cellulose and fat molecules are composed of just one or a few kinds of units, the protein molecules held many different amino acid units. Also, not all the 20 amino acid units had been discovered by the late 1800s. Yet the knowledge that protein molecules are composed of amino acid units, with structures that were easier to figure out, was a giant step forward in trying to decipher the atomic arrangement in even the smallest protein molecules.

The type of bonding between the amino acid units had not been figured out until the beginning of the twentieth century. By then the structural formulas of the known amino acids had been solved. The manner in which amino acid molecules join up to form protein molecules became known in 1907 through the work of the great German chemist Emil Fischer. But the verification of Fischer's theory in 1932, as we shall see, made use of another discovery hit upon by Schwann, the same man who was one of the founders of the cell theory, in 1835. In that year, Schwann obtained a peculiar extract from the digestive juices of humans and animals that was not acidic. Yet the extract dissolved the proteins of living tissues much more efficiently than acid solutions. Schwann had discovered pepsin, the substance in digestive juices that breaks protein molecules into amino acid molecules in the digestive process. Pepsin is itself a protein. Yet, in solution, it is an enzyme that is very effective in breaking other proteins into amino acids. We will see shortly how pepsin helped verify Fischer's theory of the bonding between amino acid molecules in proteins.

In 1901, Fischer put forth the following ideas about how protein molecules form from the simpler amino acid molecules. The structural formulas of all the amino acids contain the atomic group shown in FIG. 6-1.

6-1 Atomic group found in structures of all amino acids.

What gives each amino acid its distinct properties is the particular atomic group -R attached to the free bond on the first carbon atom of this group. The formula of the general amino acid is shown in FIG. 6-2.

6-2 General formula for amino acid.

Now consider two different amino acids given by the formulas in FIG. 6-3.

A   **B**

6-3 Two different specific amino acids. Molecules A and B combine as discussed in text.

The essence of Fischer's theory can be explained as follows: He envisioned that the two amino acid molecules in FIG. 6-3 could combine with each other if the -O-H group at the right at A latched on to one of the hydrogen atoms on the nitrogen atom at the left at B, to give a water molecule H-O-H and the two groups shown in FIG. 6-4. The free bonds on these unite to give the simple protein molecule shown in FIG. 6-5.

6-4 Two molecules resulting from reaction of amino acids of Fig. 6-3. Both A and B have free bonds that unite as discussed in text.

6-5 Protein molecule. See text for details.

This protein molecule is very simple and consists of just two amino acid units. This simple molecule can link up with another amino acid unit (FIG. 6-6) in the same fashion to give a simple protein of three amino acid units. Figure 6-7 illustrates this process.

6-6 An amino acid unit.

6-7 Formation of simple protein having three amino-acid units.

Fischer envisioned that it was possible to build up a protein chain of any length in this way. Its general form would as shown in FIG. 6-8.

The atomic groups R_1, R_2, R_3, . . ., R_N, give each amino acid unit its identity. Each of the Rs can represent any of the 20 amino acid *side chains* as they are called. Note that two factors stand out about the general protein chain. First of all one end of the chain holds the atomic group $-NH_2$ while the other end holds the atomic group $-COOH$. Secondly, notice that the groups

6-8   General form of a protein chain built up from amino acids.

$$
\begin{array}{cc}
H & H \\
C, & C \quad , \ldots \ldots, \text{ and so on,} \\
R_1 & R_2
\end{array}
$$

are connected or linked by the atomic group -COHN-, which Fischer called the *peptide linkage* or the *peptide group*. His naming of the group was probably inspired by the fact that the action of the digestive enzyme pepsin would have to break this linkage in order to yield free amino acids when proteins are digested.

From 1901 to 1907 Fischer built up simple proteins in the laboratory by adding one amino acid molecule at a time to simple chains holding two and three amino acid units, which he had also synthesized, starting from free amino acids. But his simple proteins did not behave like any of those in the living cell. That was not surprising. After all they were made to be a lot more simple in structure than those in the cell. So how could Fischer really be sure that his theory of peptide bonding applied to the proteins found in the living organism?

A strong indication that it did was provided in 1932 by other chemists through the use of pepsin solutions. When pepsin was applied to the simpler synthetic protein chains made in the laboratory, it was found that it broke them into free amino acids. This is the same effect it had on the more complex protein chains of foods in the process of digestion. In other words, Fischer had made real, though much simpler, proteins. So the manner in which amino acid molecules link up to form protein chains had been solved by the end of the nineteenth century.

A chief interest of nineteenth-century biochemists was in the role proteins played in the living cell. There was little emphasis on nucleic acids. In fact, they had not been discovered until late in the century, although the discovery of the enzyme pepsin played a role in their discovery by Miescher.

MIESCHER'S WORK WITH DNA

Frederich Miescher uncovered the substance we now call DNA in 1869. The discovery of pepsin played a key role in Miescher's work while he uncovered *nuclein*, as DNA was then called because it was obtained from the cell nucleus.

The chief objects of Miescher's research in this area were human cells from pus he obtained from bandages from a nearby hospital. He carefully washed the pus from the bandages so that the cells in the pus would not be destroyed or disrupted, and used a microscope to verify that such did not happen. In this way he could obtain intact cells that had a nucleus and cytoplasm.

He now had to separate the nuclei from the cells so that he could study them chemically. To do this he had to draw on the findings of other investigators. One such finding was that alcohol had the ability to dissolve the fatty materials in living matter, while it left the proteins and other materials unchanged. So he treated the cells obtained from the pus with alcohol to remove the fatty materials present. The other finding he drew on was the discovery of the effect of pepsin on proteins. The pepsin solution he used in these studies was obtained from the digestive juices of a pig. When he applied the pepsin preparation to the cells after they had been treated with alcohol, the proteins in them were also dissolved out, so that he had gotten a collection of free nuclei that he could analyze chemically. When he did so he found a substance that contained phosphorus, an element not found in pure proteins, and also found that the substance was very unstable and broke down chemically, unless it was prepared in a very cool place. This substance was what we today call DNA.

Miescher also found that the nuclei of several other kinds of cells contained nuclein, which came to be known as nucleic acid and showed that the substance was not affected by alcohol or pepsin. That, of course, was the basis of his procedures to isolate the substance.

Miescher's work with nuclein illustrates another important line of biochemical methods that came to fruition in the latter part of the nineteenth century. These centered around the isolation of various enzymes and other chemicals that selectively dissolve certain cell chemicals and leave others intact. These methods often supplemented microscopic investigations. We saw an example in Miescher's work. Also, in the year 1833, the French chemist Anselme Payen obtained a substance from barley that had a similar effect on starch as pepsin had on protein. He called the substance *diatase*. Diatase broke down starch into sugar much better than acid solutions did, while other such substances were found that broke fats down into their basic components, glycerol and fatty acids. In this way fats were shown to be built somewhat differently in structure than carbohydrates and proteins. Protein and carbohydrate molecules are built of chains of simple sugars and amino acids, while the structure of the general fat molecule was shown to be a little more complex.

Fat molecules contain the *glycerol unit*, derived from the glycerol molecule. The structural formula of glycerol is shown in FIG. 6-9.

6-9 Structure of glycerol.

$$
\begin{array}{c}
\text{H} \\
| \\
\text{H—C—O—H} \\
| \\
\text{H—C—O—H} \\
| \\
\text{H—C—O—H} \\
| \\
\text{H}
\end{array}
$$

The other building blocks of fat molecules are the *fatty acid units*, which are derived from the molecules of the fatty acids that are made up of a chain of carbon and hydrogen atoms of various lengths in different fatty acid molecules. These chains are attached to the atomic group shown in FIG. 6-10, known as the *carboxyl group*. Part of a general fatty acid molecule might be as shown in FIG. 6-11.

$$H—O—\overset{\overset{O}{\|}}{C}—$$

6-10 Form of the carboxyl group.

$$H—O—\overset{\overset{O}{\|}}{C}—CH_2—CH_2—CH—CH—CH_2....$$

6-11 Part of a fatty acid molecule.

The nature of a particular fatty acid depends on the specific carbon-hydrogen chain attached to the carboxyl group. Molecules of fat are formed when three fatty acid molecules unite with one of glycerol. Let the general fatty acid molecule be represented by the structural formula of FIG. 6-12, where R represents the general carbon-hydrogen chain. Then the formation of the general fat molecule can be given by the scheme of FIG. 6-13.

$$H—O—\overset{\overset{O}{\|}}{C}—R$$

6-12 General structure of fatty acid molecule.

The nature of the particular fat depends on the carbon-hydrogen groups R_1, R_2, and R_3, or the fatty acid groups comprising its molecule. So the general fat molecule was found to consist of three fatty acid units joined to a glycerol unit. Proteins and carbohydrates were found to be chains of smaller molecules.

Therefore biologists and chemists of the nineteenth century had worked out the main details of the molecular structure of fats, carbohydrates, and proteins, and did so primarily through insights gained by the

Glycerol   Three fatty acid molecules

6-13 Formation of a fat molecule.

Fat molecule Three water molecules

$$
\begin{array}{c}
\text{H} \qquad \text{O} \\
| \qquad \parallel \\
\text{H—C—O—C—R}_1 \\
| \qquad \text{O} \\
\qquad \parallel \\
\text{H—C—O—C—R}_2 \\
| \qquad \text{O} \\
\qquad \parallel \\
\text{H—C—O—C—R}_3 \\
| \\
\text{H}
\end{array}
\qquad
\begin{array}{c}
+ \qquad \text{H—O—H} \\
\\
+ \qquad \text{H—O—H} \\
\\
+ \qquad \text{H—O—H}
\end{array}
$$

6-13 Continued.

isolation of many enzymes that break these large molecules into smaller ones. This was only a beginning. At the turn of the century, many questions remained.

REMAINING QUESTIONS AND NEGLECT OF MENDEL'S WORK

Several chief accomplishments of nineteenth-century biology and biochemistry have been reviewed. First of all, newly improved light microscopes in that century made the observations of tissues and cells possible, so that the cell theory of living organisms could be confirmed. Secondly, the development of many different dyes and stains that made different cell organelles visible under the microscope shed much light on cell heredity (and heredity in general) by making precise observations of chromosome behavior in dividing cells possible. Third, the headway made in the nineteenth century in isolating the basic chemicals of cells, and deciphering their structure, was a great breakthrough. It made the progress achieved in twentieth-century biochemistry possible. The chief insight of twentieth-century biochemistry was the realization that the large molecules of life were composed of simpler molecules, which greatly simplified the problem of working out their structure. Two other milestones of that century were the discovery that proteins were the chief chemicals of life, and the finding of nucleic acid by Miescher. The first of these, though fortunate for biology as a whole, was rather unfortunate for the later development of genetics, since the nineteenth-century concept that proteins were of chief importance in the life process led to the belief that they had to be the genetic material, besides. That belief held back progress in understanding the true chemistry of heredity well into the twentieth century. Also the nineteenth-century discovery that acid solutions broke down the large molecules of life into smaller ones helped this unfortunate situation. Recall Levine's work with nucleic acids at the turn of the century. He did not realize that the strong solutions he used to separate nucleic acid from cells had the effect of breaking its large fragile molecules into smaller ones. As a result, he thought he had established the belief that the nucleic acid molecules were much simpler than those of proteins, although he was also helped in this outlook by the prevailing theory that proteins were the most complex and essential chemicals of life. Remember that Miescher had observed that nuclein, or nucleic acid, was very unstable chemically outside the cell. But, despite these misconceptions, twentieth-century biology and genetics owes much to these nineteenth-century developments.

Nineteenth-century insights into these areas had its limitations. First, the specific functions of many organelles in the cell remained unclear. Also, their specific chemical compositions could not easily be probed. It was difficult enough to isolate a given kind of cell with the methods available at the time, let alone large numbers of a given kind of organelle to analyze its chemical makeup. Miescher's ability to isolate certain cell nuclei was a fortunate exception. The task often depended on the existence of various solvents (like pepsin or alcohol in his case) that would selectively dissolve all organelles except the one sought. That was asking much at the time, and is not really possible today. All organelles often contain some protein (or some other constituent) that is affected by a wide range of solvents. So most solvents, both enzyme solutions and others, deform most cell organelles to some extent. This makes their separation from one another more difficult. Even if such separation was possible it would therefore not give a true picture of the organelle's composition, since some of its chemical constituents are decomposed or removed through the use of most solvents.

A similar situation existed in the nineteenth century with regard to knowledge of the structure of specific proteins and other chemicals of life. For example, chemists of the time discovered many of the amino acids composing proteins, and also realized that proteins must be built of simple amino acid molecules. They also determined the structural formulas of many amino acids and simple sugars near the end of the century. Yet, in fact, two crucial problems still faced them. One was the question of what specific proteins, fats, and carbohydrates, and so on, composed various cell organelles, while the other centered around proteins, in particular, and, later, in the twentieth century, on nucleic acids. How could the different kinds of amino acids in a given protein molecule, the number of each, and their arrangement along the protein chain be determined? Sometimes twentieth-century biologists and biochemists were fortunate enough to be able to isolate a given protein from an organism. They could then dissolve the protein in acid or pepsin solution. But all their treatments could do was to break the protein into its component amino acids. If the chemist was fortunate, he could manage to separate one or two specific amino acids out of the resulting mixture by techniques available. But most of the amino acids would stay mixed up. No adequate chemical methods existed to separate all the components of the mixture from one another. There, however, was another serious handicap that hampered all early biochemical studies. That was that most important biochemicals could be isolated from cells only in very small amounts, if at all. So the nineteenth-century chemist was faced with two severe problems: he had to try to analyze a very complex mixture for its precise composition, and this mixture came in very small amounts. Clearly the problem was beyond the chemical technology of the times. It was not solved until well into the twentieth century.

Adding to these limitations the fact that precise research on heredity had not been conducted during the nineteenth century, it is easy to understand how Mendel's work was neglected then. The idea that a biological phenomenon obeyed mathematical laws did not attract the attention of many contemporary biologists. In the case of heredity, Mendel had established that inheritance obeys statistical laws. Biology had been a descriptive science that primarily concerned itself with the structure and classification of organisms, and had little in common with physics and chemistry, the exact sciences, as far as the application of mathematics was concerned. Biological phenomena were thought to be too complex to obey simple mathematical laws. That was one of the chief reasons why many life scientists did not take Mendel's original papers seriously. At most, the scientific community of the time received his findings with interest, but little more. Many biologists, in fact, could not follow many of the arguments put forth in his papers because they were expressed in equations from the field of mathematics known as *probability theory*. The use of such methods were, for the most part, foreign to biological science in the nineteenth century. There were exceptions among individual researchers, but such cases were not the rule. Often the theories were not nearly as productive as Mendel's in the area of heredity. Such efforts merely went down as attempts that were original in approach. But they did not have the genius or originality behind them that were found in Mendel's studies of inheritance, when sections of his papers were discovered at the turn of the century.

Mathematics was not the only barrier to immediate acceptance of the gene theory. There was also his belief that traits were regulated by particulate factors. At the same time, he had no clear idea of what these particle-like factors might be. Cell organelles were just beginning to be observed by some biologists. However, in the 1860s, the time of Mendel's work, the technique of staining cell organelles to make them stand out against the transparent background under the microscope had not yet been fully developed. It was the specialty of only a minority of biologists of the time. So the question of whether or not cells contained smaller parts was only a matter of supposition to many experts. Thus Mendel's idea, though it accounted for the observed facts quite well in many instances, lacked real experimental confirmation.

There was a third factor that cannot be passed over lightly in understanding why Mendel's work was not appreciated for its originality in the nineteenth century. This matter centers around the fact that he used the methods of statistics. His approach must have gone something like this: He assumed that each of the two aspects of each simple trait was regulated by some kind of small physical body, today called a gene, that is carried to the seeds in the germ cells of the plant or animal. Each germ cell in his first generation hybrid plants held one gene for a given aspect of a trait. Also the two kinds of germ cells that held the genes for the two aspects for each trait were seen as produced by each parent plant in equal numbers. At fertilization, these germ cells unite at random to produce the first cell, or *zygote*, of each second generation plant in the seed that gave rise to the plant. By all indications, he made these assumptions before he conducted his experiments. He then predicted how many offspring of each type should arise in his second generations on basis of them. Thus, for example, when he crossed the tall first generation hybrids with each other, he deduced that three out of every four of their offspring should be tall and one out of every four dwarfed. Yet there was one complication at this stage: these were only statistical predictions. They did not mean that if one selected four seeds at random from a pool of seeds of the second generation plants that precisely three of them would sprout into tall plants and one into a dwarfed plant. In such a small sample of second generation seeds, only one seed might give a tall plant and the other three dwarfed ones. Then what precisely did Mendel's statistical predictions mean? All they meant in his first experiments, for instance, is that the 3 to 1 ratio of types of the two offspring was only obtained approximately as the total number of seeds became very large. In mathematical terms, the 3 to 1 ratio was a limiting ratio that would be gotten if one imagined the number of seeds giving the second generation plants getting larger and larger without bound. If one gets only small numbers of seeds out of a cross, the number of tall and dwarfed plants will often deviate greatly from the 3 to 1 ratio. That is because Mendel's theory made statistical, not exact, predictions.

But, then, one big problem must have confronted Mendel. How large a number of seeds, or offspring, was large enough to test his theory experimentally? To answer this question, one had to be well grounded in mathematical ideas and in probability theory. Here Mendel was fortunate. His monastic education included much training in physics and mathematics, and therefore, he must have known that he would have to raise at least thousands of offspring in each second generation to have any chance of testing his theory adequately. That is just what his experiments in the monastery garden were designed to do. He must have gotten at least thousands of seeds in the trials of his experiments.

The question still arises: is thousands of offspring a large enough herd to test the theory? Mendel claims to have verified the theory to his own satisfaction under such conditions, although some researchers today have claimed that his results seem too good to be true compared to results gotten through modern breeding experiments. Mendel did not have all the conveniences of the modern breeding laboratory, and had to do the best he could with the monastery garden. What he probably did was to select the cases that seemed to confirm the gene theory, and tried to account for deviations from the predictions of the theory through other factors such as climatic conditions, possible counting errors, and other items that might have influenced the experimental results which he had no way of knowing about. But it does seem likely that he found many cases that seemed to confirm the theory under the conditions prevailing in the monastery garden.

All this does not change the fact that Mendel had arrived at a very productive theory of heredity. Time has shown that his model has triumphed over all others.

The most popular model for explaining heredity at Mendel's time was the *blending theory*. According to this theory, the life fluids of the parent organisms mixed in the offspring, so that the traits of the offspring would be a mixing, or blending, of the traits of the parent organisms. Thus if a plant with red flowers is crossed with one having white flowers, the offspring should have pink flowers, pink being a blending of red and white. Similarly if a tall plant is crossed with a dwarfed one, the offspring plants should have a height somewhere between the heights of the parent plants in this theory of heredity. The fact that many nineteenth-century experts held to this theory was not blindness on their part. Many examples like those just given seemed to show the validity of the blending theory. In fact, Mendel studied a case like the first example: one in which a plant with pink flowers arose from parents having red and white flowers. But it was clear that the blending theory could not explain his results with pea plants. For example, the dominant traits came through unblemished in the first generation hybrids in all cases. Blending theory had no explanation for this. Worse yet, it had no way whatever of accounting for the reoccurrence of the recessive traits in his second generations after being absent in the first generation hybrids. But what could be made of his experiments with the red and white flowered plants? Did he have to admit that blending theory might be valid after all, at least in some cases?

Not at all. His handling of this case well demonstrated the power of the new mathematical methods he brought to the study of heredity. The approximate proportions of the red, white, and pink flowered plants in a second generation of a breeding experiment with the plants showed beyond a doubt the superiority of the gene theory over blending theory even in this case. He started out with two sets of plants. One of them was pure in the trait of red flower color. The other was pure in the trait of white flower color. Mendel then crossed the red flowered plants with the white flowered ones, and found that all the first generation plants had pink flowers, a fact that could in itself be accounted for by blending theory. At this point Mendel made another assumption. What if the gene for red flower color R interacted with that for white flower color W in the gene pair for flower color, so that all the first generation plants having both genes had pink flowers? Then neither gene would be dominant or recessive. The effects of the two genes would blend in the offspring, although the explanation of this was not to be found in the mixing of life fluids. How could he test this supposition further? All he had to do was to cross the pink flowered offspring with one another. Now the first generation offspring had the genetic composition RW in the gene pair for flower color. Therefore, by Mendel's gene theory, they would produce germ cells with the compositions $\boxed{R}$ and $\boxed{W}$ in equal numbers. When the first generation pink flowered plants are crossed, these unite at random according to the scheme shown in FIG. 6-14.

Parent 1		Parent 2		
R	+	R	⟶	RR
R	+	W	⟶	RW
W	+	R	⟶	WR
W	+	W	⟶	WW

6-14   Crossing of red (R) and white (W) flowers.

From this scheme Mendel could predict the proportions of these different kinds of offspring in the second generation. The kind that had the genetic composition [RR] had red flowers and should make up about one fourth of the second generation. The type having the genetic composition [RW] had pink flowers and should make up about half of the second generation made up of a large number of plants. The last type of the genetic composition [WW] had white flowers and should make up about one fourth of the second generation. In other words, Mendel could predict that about one fourth of the second generation plants should have red flowers. Half should have pink flowers. The last fourth should have white flowers. These are approximately the proportions of the three kinds of offspring Mendel claimed to have found in the second generation of plants. If the blending theory held, all generations of plants after the first should have had pink flowers, while the reoccurrence of plants with red and white flowers in that generation could not be accounted for in blending theory.

This example best brings out how two new innovations brought to the study of heredity, one of which was the systematic experiment making use of simple, easily observed traits, while the second was something new not only to the study of heredity, but also to biology as a whole. This was Mendel's use of mathematical schemes to test theories. This approach had been characteristic of sciences like physics and physical chemistry but had not been used in biology, the study of life, to any significant extent. Mendel's mode of attack seemed to have been to assume the truth of a theory cloaked in a mathematical framework, and then draw conclusions from it that should be found to hold true experimentally if the theory is valid. This is the same powerful method that had been being used by physicists in their work for over two centuries. Mendel's approach was no doubt influenced by his monastic education in physics and mathematics.

Although the reoccurrence of traits unblemished in the first and later generations spoke clearly enough for the truth of the gene theory, the mathematical aspect of the theory gave it an air of exactness. So, despite obscurity of the gene theory in the nineteenth century, it must be classed as one of the chief developments in biology and its methods. It posed many unanswered questions when it was discovered at the end of that century. Some of these questions were already being asked by others during the last two decades of the century.

One chief problem near the end of the nineteenth century centered around the significance of certain thread-like entities seen in cell nuclei. The observation of these bodies went back to the last part of the 1870s. Then the German cytologist Walther Flemming applied the newly developed stains to cell nuclei, and, when he applied them to stable cells, he found colored dots composed of some material that was part of the cell nucleus. The dyes he used killed the cells he studied. But he also decided to apply them to cells of growing tissue at different stages of the tissue's development in the organism. In this way he caught various cells at different stages of division. He observed that as the cell was about to divide, the colored spots formed themselves into colored, threadlike bodies that seemed to duplicate themselves just before cell division. As the cell divided, half of them went to one daughter cell and half to the other. Flemming named the material making up the thread-like bodies *chromatin*, because of its ability to become brightly colored in the presence of the dyes he used. Flemming discovered what we today call chromosomes and their duplication, though he did not call them chromosomes. He also discovered the fact that they were divided equally among the daughter cells. He observed these happenings under the newly improved microscopes. This was one early example of how the different kinds of biological methods aid each other.

In 1887, the Belgian cytologist Edward van Beneden followed up these discoveries, and found additional properties of the thread-like bodies in cell nuclei. First of all, their number before cell division seemed to be the same in all cells of a given organism. Secondly, each organism seemed to have a fixed number of them in the nuclei of their cells which depended only on the organism. He was also able to observe them in the germ cells of organisms, which seemed to have half the normal number of the thread-like bodies in their nuclei.

Such discoveries made with the new microscopes and staining dyes had great significance when Mendel's work came to light at the start of the twentieth century. By then the thread-like bodies came to be called chromosomes. Little did Flemming and van Beneden seem to be aware that the chromosomes they studied were the carriers of heredity in all organisms, although they no doubt wondered what the function of these strange organelles could be. It is mind-boggling for us today to realize that Mendel's papers (postulating particle-like factors as carriers of traits) had been published for over a decade while the two investigators were using new technologies to observe bodies in cell nuclei that behaved like Mendel's genes in every apparent way. This all goes to show how revolutionary Mendel's theories and approaches were to biology of that time.

The gene was basically a chemical entity. At that time, serious study of the chemistry of cells was just getting under way. The big question of how the chemical processes in cells were actually carried out had not been touched upon, since very few pure proteins (and other life chemicals) had been isolated. Also, the isolation of specific cell organelles in large numbers was, for the most part, beyond the experimental technology of the nineteenth century. So one of the first problems that had to be solved was that of how to obtain larger collections of a given type of organelle in order to study their chemistry. That was tackled in the twentieth century. There was also the need to isolate more cell chemicals in more pure form. But that was only part of the task. The chemicals then had to be analyzed for their specific molecular structures in terms of the simpler building blocks that chemists of the last century had shown them to be composed of the amino acids, fatty acids, and simple sugars. The development and perfection of experimental techniques to do these things took up the first half of the twentieth century. With their use, biochemists and geneticists were able to answer the additional questions posed about the genetic code at the end of Part I of this book.

Chapter 7

Early twentieth century methods

*T*he twentieth century opened with the discovery of Mendel's work. This breakthrough got experts pondering what the basis of heredity might be. By 1900, the time of this event, chromosomes had been seen in cell nuclei, and their behavior during cell division was well documented by many microscopic studies. The first man to propose the idea that chromosomes behaved like the physical heredity factors envisioned in Mendel's theory was the American scientist Walter S. Sutton. He was a cytologist and therefore aware of late nineteenth century findings made with the newly improved methods of staining and microscopic studies with regard to chromosomes and their behavior. Sutton noticed that chromosomes were passed to daughter cells during cell division in the same random way Mendel's genes were supposed to be passed to offspring. They also seemed to be the only organelles that retained their identity during cell division, and they existed in all cells of the organism. Their number was fixed for each organism. All this made them very similar to the genes of the gene theory. So very quickly chromosomes became the carriers of genes in the opinions of many scientists.

PRELIMINARY DEVELOPMENTS

In the first half of the twentieth century, not much headway was made in understanding the chemical structure of chromosomes or the chromatin that comprised them. That they contained nucleic acid and protein was known. However, the real nature of nucleic acids and the protein that accompanied them in chromatin was unclear at the start of the century. The biochemist Phoebus Levine began research in 1910 that eventually showed that DNA molecules were composed of nucleotides. Each nucleotide he postulated was composed of a phosphate group, a sugar group, and a purine or pyrimidine group linked together. But his technique of isolating DNA from cells was only a little better than Miescher's. Miescher noted that nuclein (or DNA) was unstable chemically, and Levine made use of strong acid and alkaline solutions in separating nuclein from cell nuclei. As was well known in the nineteenth century, such solutions tended to disrupt molecules of even more stable biochemicals like starches and fats. The strong solutions Levine used broke the DNA molecules into much smaller ones. Through chemical procedures of the early 1900s, Levine could determine the weight of these molecules. Their weights seemed to be equal to that of molecules made of four deoxyribose nucleotides joined together. Thus arose the four-nucleotide theory of DNA structure. Again, it is strange that Levine did not seem aware that the solutions he used might have broken larger biomolecules into smaller ones, since, after all, Miescher had noted that nuclein was very unstable, while nineteenth-century chemists had shown that acid solutions broke large organic molecules into smaller ones.

Here again, the only explanation seems to be related to the central role ascribed to proteins in the concepts of the time. Proteins seemed to be the only really complex molecules in the living cell. This belief lasted far into the present century as far as the role of nucleic acids was concerned, and it is likely that Levine found nothing unusual in his finding that DNA had relatively small molecules. If chromosomes contained the newly discovered carriers of heredity, it seemed a natural conclusion that its protein part played the chief role in the process of heredity. Nucleic acids held little weight in the thoughts of biochemists and geneticists until well into the 1940s.

During the first three decades of the twentieth century, biochemical methods developed rather slowly. Little real progress was made in separating specific proteins from living matter in a systematic way, and their isolation was still largely a matter of accident. In this way, there had been little advance over nineteenth century efforts.

EARLY BREEDING EXPERIMENTS

These are the main reasons why studies of heredity in the early decades of the present century stressed breeding experiments rather than the approach of chemistry. Of course, all the various methods of genetics and biochemistry are equally essential to the understanding of heredity and how it works. This is not to say that breeding experiments are superior to other methods. It only means that such experiments combined with microscopic studies were the only real paths open to geneticists in the early decades of the twentieth century. So, at that time, geneticists started looking for organisms that were more suited for breeding experiments than Mendel's pea plants had been, and, in the year 1907, the small fruit fly *Drosophila melanogaster* came to attention of the great American geneticist Thomas Hunt Morgan. The fly had a characteristic that made it excellent for both breeding experiments and microscopic studies. Here again is a historic example of two important experimental methods aiding each other. The salivary glands of the fly were found to contain cells that had very large chromosomes. They are about 100 times as thick as chromosomes in most other cells and organisms. Thus these chromosomes could be seen and followed much more easily under the microscope than most others. When the salivary glands were stained to make the large chromosomes visible, each such chromosome was found to contain many clearly visible bands of various sizes along its length. These bands held many surprises for geneticists of the early twentieth century.

It soon became apparent that the bands on the large chromosomes were sites of genes for various traits of the fly. When Morgan found a mutation in a fly, he often examined its salivary gland chromosomes under the microscope. This is how he often found some deformation or irregularity in a band of one of the mutated fly's large chromosomes. The irregularity was absent in the same band of a normal fly without the mutation. What these observations clearly indicated was that the gene that governs the normal form of the mutated trait of the fly is contained in the band of the chromosome that showed the irregularity in the mutated fly. The study of these chromosomal bands along with numerical results of breeding experiments helped Morgan map the four chromosomes of the fruit fly for their various genes.

Microscopic studies had shown that the fly had four sets of chromosomes. In the fly, each chromosome set except one held two chromosomes of the same length, thickness, and band structure. The chromosomes of these three sets are of different lengths. Each of the two chromosomes in any one of the three pairs are of the same length. The fourth set, or pair, was an exception to this pattern, because one chromosome of the pair was much longer than the other. The shorter chromosome of the pair seemed to occur in male flies only. The same chromosome pair in female flies was seen to be made up of two of the longer chromosomes that are identical. The longer chromosome of the pair in male flies became known as the *x chromosome*, while the shorter one became known as the *y chromosome*. The same phenomenon had been observed in the chromosomes of other organisms in the 1890s, and in both Morgan's studies and these earlier ones, female organisms were always found to have two X chromosomes in the pair. Male organisms of the same species were always observed to have an X chromosome and a shorter Y chromosome in the same pair. What this meant was quite clear.

The X and Y chromosomes seemed to be the chromosomes that determined the sex of the organism. One pair of chromosomes seemed to serve this function in a lot of organisms. If the pair held two X chromosomes, the organism was always observed to be female. If the pair held an X and Y chromosome, the organism was always observed to be a male. Morgan's studies of Drosophila salivary gland chromosomes further confirmed this theory.

His observation of the fly's sex chromosomes shed light on another fortunate occurrence he had take note of at the beginning of his genetic studies. This was the appearance of the one white-eyed fly in one of his first experiments. That, recall, was a fortunate incident for him. It provided him with the first mutation he could follow from parent to offspring in the flies. But there was a strange aspect to the mutation. For one thing, the white-eyed fly was a male. That, in itself, does not mean anything spectacular. What was surprising was what followed. Recall that Morgan mated the white-eyed fly to a regular red-eyed female fly, and all flies coming out of the cross were normal red-eyed ones, both male and female. That seemed to show that the trait of white eye color was recessive. But when he crossed two of the first generation red-eyed flies, both red-eyed and white-eyed flies were found among their offspring. One striking fact stood out. All the white-eyed flies were males like the original white-eyed one. All the females were red-eyed. No red-eyed female ever turned up. Why? Clearly the traits of eye color and sex were linked in the fly. However, on basis of microscopic studies of the sex chromosomes of the fly, Morgan could say more than that.

He had established that the gene for eye color must be located on the sex chromosomes of the fly. These were strange chromosomes; they were of different lengths. The female chromosome is longer than the male sex chromosome. The female sex chromosome, in other words, had a stretch of genes that the male sex chromosome did not have. So the first question was whether the eye color gene exists in the stretch of genes the two sex chromosomes have in common, or in the part of the female sex chromosome the male sex chromosome does not have. If the first of these hypotheses is correct, then both chromosomes of the pair hold an eye color gene, while, if the second is correct, only the female sex chromosome holds the eye color gene. Morgan could get an idea about which of these is the case by going back to the white-eyed, male fly that had turned up in his earlier experiment. This fly would have an X chromosome and a Y chromosome, since it was a male. Imagine that the eye color genes are on the portions of the two sex chromosomes the pair have in common. Now natural mutations occur at germ cell formation. Our assumption would mean that the spontaneous gene mutation, that gave rise to the unnatural gene for white eye color, would have had to occur simultaneously in the two germ cells that gave rise to the zygote of the white-eyed fly. That would be highly unlikely, though not impossible. The likelihood of both sex chromosomes holding the eye color gene was further diminished by the fact that not one white-eyed female fly was seen in the experiment. All evidence spoke for the belief that the eye color gene was held only by the X chromosome.

If that is the case, the results of Morgan's experiments are easy to understand. The first white-eyed fly had inherited an X chromosome from its female parent that held the recessive gene for white eye color. Since that gene exists only on the X chromosome, there was no corresponding gene for the dominant trait of red eye color on the fly's Y chromosome to block its expression in the fly. The female fly mated to the white-eyed fly had two X chromosomes, each having the normal gene for red eye color. So its germ cells had only X chromosomes with the gene for red eye color. But the white-eyed male had two kinds of germ cells. One kind held the X chromosome with the recessive gene for white eye color. The other kind held the Y chromosome that held no gene for eye color. Call the dominant gene for red eye color R, and the recessive one for white eye color r. The possible results of the mating between the normal female fly and the white-eyed male fly can be represented by FIG. 7-1A. $\overline{X}_R$ is the germ cell holding the X chromosome of the normal female fly. $\overline{X}_r$ is the germ cell holding the X chromosome of the male that has the recessive gene r. $\overline{Y}$ is just the germ cell holding the male sex chromosome. Cases 1 and 2 give normal female flies with red eyes. Cases 3 and 4 give normal red-eyed male flies. So all the first generation flies are red-eyed, as Morgan found was the case.

Female parent

	$\overline{X}_R$	X_R	
$\overline{X}_r$	**1** $\overline{X}_R$ $\overline{X}_r$	**2** $\overline{X}_R$ $\overline{X}_r$	**A**
$\overline{Y}$	**3** $\overline{X}_R$ $\overline{Y}$	**4** $\overline{X}_R$ $\overline{Y}$	

Male parent

Female parent

	$\overline{X}_R$	$\overline{X}_r$	
$\overline{X}_r$	**1** $\overline{X}_R$ $\overline{X}_R$	**2** $\overline{X}_R$ $\overline{X}_r$	**B**
$\overline{Y}$	**3** $\overline{X}_R$ $\overline{Y}$	**4** $\overline{X}_r$ $\overline{Y}$	

Male parent

7-1 Dominant and recessive genes for eye color. See text for details.

However, note that all the first generation female flies had the recessive gene for white eye color r on one of their X chromosomes. So they produce two kinds of germ cells, which can again be given by the symbols $\overline{X}_R$ and $\overline{X}_r$. But the male first generation flies produce the germ cells $\overline{X}_R$ and $\overline{Y}$. Figure 7-1B shows the mating of two of the first generation flies. Cases 1 and 2 give female flies with red eyes. Case 3 gives a male fly with red eyes, while case 4 gives a white-eyed male fly. So the second generation should be made up of red-eyed males and females, as well as white-eyed males. That is what Morgan found in that generation.

This example shows how microscopic investigation helped clarify one of Morgan's important findings in his breeding experiments.

For the first three decades of the twentieth century, biochemical and genetic technology had made only one main stride over nineteenth-century methods: the breeding experiment. The value of this kind of experiment had become apparent with the discovery of Mendel's work and the fruit fly *Drosophila melanogaster*. Some headway had been made to dispel some of the mystery as to where Mendel's physical factors of heredity were located in the cell. Morgan's work with the fruit fly verified Sutton's hypothesis that the genes are located on the chromosomes of cell nuclei. It also showed that genes are arranged in linear order on the chromosome. So, by the early 1930s, the gene was seen as a segment of the chromosome. Yet chemical techniques of probing the gene had not advanced much over those of the nineteenth century. Only a few pure proteins could be isolated from cells. There was no way to obtain many specific proteins in large enough quantities to analyze them chemically for their amino acid composition. This was particularly true of enzymes. Enzymes occurred in very small amounts in the cell. Few of them could be gotten in pure form. But some important strides had been made in the early twentieth century in this area.

ENZYMES

The theory of enzymes as organic catalysts goes back to the early nineteenth century. The action of enzymes was then seen as a specific case of the process of catalysis. Even eighteenth-century chemists had observed that many chemical changes were hastened by the presence of certain substances that were not changed themselves in the change. In 1812, the Russian chemist Gottlieb Sigismund Kirchhoff was the first to note that a heated solution of an inorganic acid could break down starch into glucose, although the acids used were common inorganic ones that had not been obtained from living cells. Then, in 1816, the great English chemist Humphry Davy had shown that the combustion reaction in which alcohol combines with oxygen could be made to take place at ordinary temperatures in the presence of the metal platinum. In the absence of platinum, alcohol does not burn unless it is heated to much higher temperatures. The platinum was not changed itself in the reaction; it acted as a catalyst. But all these catalysts were inorganic ones. After these discoveries, the idea that living matter might contain organic catalysts that enabled the cell to carry out many chemical reactions at a low temperature began to seem reasonable to some chemists of the time. Some of these catalysts like diatase and pepsin had been isolated. But they had not been gotten in pure form from the organic solutions that contained them. Investigators merely assumed that the action of such solutions on starch and other substances from the living organism was due to some ingredient in the solution that broke these materials down. The enzymes themselves had not been separated from the solutions. They were like a ghost in the solutions that had a puzzling effect on substances from living matter.

This situation had not changed by the early twentieth century, despite Fischer's success in putting together simple peptide chains in the laboratory. It was not even certain in 1902 whether or not enzymes were proteins. It was suspected that they were; heating an enzyme solution destroyed its ability to function. Proteins had this same characteristic. Therefore, since no enzymes had been isolated in pure form, the belief that they were protein in nature was just theory at the turn of the century.

In 1904, evidence was uncovered that implied that enzymes were not pure proteins in all cases, although most of their molecular structure was thought to be so. In that year, the English biochemist Arthur Harden shed some light on the question. His experimental technique made use of another device that has had some importance in biochemical methods of the past and present: the semipermeable membrane. A semipermeable membrane is a very thin sheet of some material like parchment that is full of tiny microscopic holes or pores that are too small to be seen with the naked eye. Say we have a sheet S of such material that separates a container C into two parts or compartments A and B, each of which can be filled with two different liquids as shown in FIG. 7-2. To grasp the significance of Harden's work, we must understand exactly what a semipermeable membrane does. Say that compartment B is filled with pure water, while A is filled with water in which a large amount of table salt has been dissolved. Originally compartment A holds salt solution, while B holds pure water. After a certain time passes, the liquid in both A and B will be found to contain salt. There is only one explanation for this phenomenon. Recall that molecules are always in motion. Apparently those of salt are small enough to pass through the small pores in the semipermeable sheet S and enter compartment B. After a while, the solution in both A and B will contain equal amounts of salt. The molecules of most simple

Container C

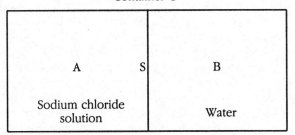

7-2 The use of the semipermeable membrane in biochemical methods.

sugars and table sugar are also small enough to pass through the pores in such a membrane. The passage of substances in solution through such membranes was first studied intensively in the middle of the nineteenth century. It was found that molecules of simple salts and sugars could pass through such membranes, but those of a protein such as gelatin could not, while the same was found to be true for proteins in general. Apparently proteins had molecules too large to pass through the pores in the membrane. Thus they could not get through the membrane.

Harden's work in 1904 made use of a solution obtained from yeast. The yeast solution had the ability to break glucose into alcohol at room temperature. So it was assumed that the solution obtained from yeast must contain an enzyme for breaking down sugar into alcohol and other simple substances. The fact that yeast could do this was known before 1904. Harden made use of the fact in his work. He filled a bag made of a semipermeable substance with the solution obtained from yeast. He then immersed the bag in water for some time, and found by chemical tests that many substances in the yeast solution had passed from the solution into the water. That was not all. Before the substances passed into the water, the yeast solution could break down sugar into alcohol. However, after they had passed into the water through the membrane, he found that the solution left in the bag could not change sugar into alcohol any longer. The enzyme that was responsible for the change seemed to have lost its power to do so after certain substances had passed out of the solution. The enzyme itself could not have passed out of the solution. Two observations spoke against this. First, it had been shown that substances that pass through semipermeable membranes pass through only to the extent that their concentration—the number of molecules of the substance per unit of volume of the solution—became the same on both sides of the membrane. In other words, some of any substance that had passed through the bag should have remained in the bag. So some of the enzyme would still have to be in the solution in the bag, though it would be very small, since a large volume of water had surrounded the bag. The solution in the bag should still have had some ability to break down sugar, but it did not. Could all of the enzyme have somehow passed into the water outside the bag? A second observation spoke against this: The water solution outside the bag did not have any ability to break down sugar either.

There was only one answer left. Part of the enzyme molecule must have become dislodged from a much larger part of the molecule. This smaller part of the molecule was small enough to pass through the pores in the bag, while the larger part was too large to pass through them. Thus the enzyme molecule in the yeast extract responsible for the breakdown of sugar must be composed of two parts: a large molecule and a much smaller one bound together when the enzyme was active at its function. This was shown to be the case in another part of Harden's experiment. In it he mixed the solution left in the bag with the water solution outside. The resulting solution had marked ability to break down sugar. Apparently, the two parts of the enzyme molecule had recombined to give the complete enzyme molecule when the two solutions, one containing each, were mixed. The small part of the molecule was obviously necessary for proper functioning of the enzyme.

What was the nature of the small portion of the enzyme that was necessary for its operation? In 1904, that was quite a loaded question. After all, the true chemical nature of enzymes had not yet been established. The fact that the enzyme in the yeast solution contained a large part that did not pass through the membrane, strengthened the view that enzymes were protein in nature. Proteins had the same property. However, there did not yet seem to be any real proof that enzymes were proteins.

That began to change around 1926. In that year the American biochemist James Summer was doing research on an enzyme called *urease*, and he had gotten a solution containing the enzyme from jack beans. Urease breaks urea into carbon dioxide and ammonia at room or body temperature. In the process of separating the urease solution from its source, the jack beans, Summer noticed that sometimes small solid crystals separated out of the solution. He collected them and redissolved them to get another solution, which had the ability to break urea into ammonia and carbon dioxide. A sound conclusion was that the enzyme urease must be contained in the crystals. Summer put the crystals through diverse types of chemical analysis, but no matter what he did to them, he found only a material having all the chemical and physical characteristics of the original crystals that had the ability to break urea into ammonia and carbon dioxide. The only one conclusion that could be drawn from this was that the crystals must be made of pure urease. Urease was thus the first enzyme to be isolated in a solid form. This is a noted example of a case in which a fortunate chance incident, the appearance of the white crystals, played a major role in the progress of a given area of science.

The crystals of urease were shown to be pure protein. All this was only a beginning, however. Additional enzymes were separated from extracts (gotten from living organisms) in solid form. Beginning in 1930, crystals of the enzymes pepsin, trypsin, and chymotrypsin were obtained in the laboratory, the last two of which are also enzymes that break proteins into amino acids. They, too, were shown to be proteins. Other enzymes were also gotten in the form of crystals and analyzed; most were shown to be proteins. That is, for the most part, because of Harden's discovery in 1904.

In the 1930s, progress was made in isolating the nonprotein portions of the class of proteins of which Harden's had been an example, and the structural formulas of many of the smaller nonprotein portions of such proteins (called conjugated proteins) were figured out. The portion of the protein molecule that easily diffused through a semipermeable membrane, though not composed of amino acids, was found to be necessary to the functioning of these enzymes. The class of small molecules that formed a small part of larger enzyme molecules in this way became known as *coenzymes*. It was determined that the vitamins taken in through the diet supplied many of the coenzymes. The organism cannot make these nonprotein portions of its enzymes on its own, and needs to take them in through the diet.

THE GENE QUESTION

By 1930, enzymes were established as proteins. But there was still no great improvement in biochemical and cytological methods over those of the nineteenth century. Cell organelles could not yet be isolated in large numbers. Also specific enzymes and other proteins could not be gotten to any great extent. All evidence for the existence of many of them still remained indirect and rested on fortunate incidents in the case of others, while little headway was made in understanding the chemistry of genes and how it regulated the process of life. The discovery that enzymes are proteins was unfortunate in a way for the probing of the chemical nature of the gene: it further strengthened the belief that the genetic material could only be a protein. The fact that the catalysts of the life process are protein added weight to the conviction that proteins are of first importance in the chemistry of life. Thus it seemed all the more unlikely that the chief regulators of life, the genes, could be anything except proteins.

Many protein molecules had been established to hold parts that are not made of amino acids. So the fact that chromatin of the nucleus contained both proteins and nucleic acids was not very troublesome for believers in the protein concept of the gene. Also Levine thought he had shown that nucleic acid molecules were small ones alongside those of most proteins. He did not seem to appreciate the instability of nucleic acids as Miescher had. Thus it could only be concluded that nucleic acids played a secondary role in the chemistry of chromatin.

However, even in the 1920s, new methods of investigation were being developed that would eventually show that such was not the case. One of these was the invention of a device called the *centrifuge*. Another was the study of *X-ray diffraction* by crystals of substances.

THE ULTRACENTRIFUGE

Viewing various cell organelles through the microscope, though it shed some light on various theories of genetics, had some limitations. First of all, it gave no real information about the chemical composition of the organelles. If cytologists and biochemists were to learn more about the purpose of each organelle in the overall machinery of the cell, the chemical composition of organelles was important knowledge. After all, the cell organelles are the most fundamental components of the life process. The next more fundamental components of the life process. The next more fundamental level is the biomolecule. And at that level, the only relevant concern is the chemical interaction between the molecules, which is a chief quest of modern molecular biology.

Secondly, while the microscope could give some knowledge about the shapes and relative sizes of different cell organelles in the cytoplasm and nucleus, a rough idea of how many of each organelle exist in the cell, and a little information about the purpose of some organelles, such knowledge did not say much about the function of each organelle in the plan of the cell, nor about how all of them cooperated as a whole to give a living dynamic cell. So the microscope had to be supplemented by other modes of cytological investigation.

If any headway was to be made in working out the chemical composition of each organelle, a method of isolating a given kind of organelle in large numbers was needed. It had been easy in the 1920s to get a mixture of different organelles from a given kind of cell. One method of doing this was simple. It was only necessary to take a given kind of tissue and grind it up in some kind of vessel containing a liquid such as a concentrated sugar solution. The grinding process crushed the individual cells. The cells then spilled their contents into the liquid. The result was a mixture of cell organelles, for the most part, and disrupted cell membranes. But if the mixture in which this process is carried out is water, the different cell organelles tend to clump together in it. That would make it more difficult to separate them. In a sugar-water mixture, the various organelles stay separated from one another. The sugar that is usually used is sucrose, or table sugar. But how could the components of such a mixture be separated from each other? This was quite a problem in 1920.

A technical breakthrough was made in 1923 that was to have great bearing on the solution of this important problem. In that year, a Swedish chemist came up with a brilliant idea. In short, he invented a device which is known as the *ultracentrifuge*, or simply the *centrifuge*. Such a device made use of the fact that if a test tube containing a mixture of different organelles in sugar solution is spun rapidly around an axis perpendicular to its length, the different organelles move down the solution in the tube at different rates. This results in their separation from each other because each tends to settle in a different part of the tube.

The physical principles can be understood by considering something experienced on a moving merry-go-round. Imagine that the merry-go-round is spinning quite fast. Anyone on the merry-go-round had better hold on to the device tightly. If they do not, they will be thrown off almost instantly. The further a person is situated from the center of the merry-go-round, the greater will be the force tending to throw the person off. Also, at a given distance from the center, the heavier the person is, the greater this force seems to be. A full grown person will be throw off easier than a small child at a given speed of revolution of the merry-go-round. So there seems to be three factors affecting the force with which a person is thrown off the merry-go-round if not holding on to the device tightly: the rate at which the device is spinning, the weight of the person, and the distance of the person from the center of the device.

Now imagine a merry-go-round made up of a number of circular rings of seats. Each circle of seats is situated at a different distance from the center. Then people seated in any one circular seating ring are at the same distance from the center of the merry-go-round. Say the merry-go-round is spinning at a constant speed. Imagine also that this merry-go-round has a wall built around its outer edge to prevent people from falling off if they let go of their seats and stand up in the spaces between the seats. Also imagine that there are many people of different weights and sizes seated all over the merry-go-round. If they all stand up in the spaces between their seats, each will seem to feel a force throwing them toward the outer edge or wall surrounding the device. The farther persons of about the same weight are from the center of the merry-go-round, the greater the force. Also, the heavier persons in each circular row of seats will feel themselves thrown against the wall harder than the lighter ones. Also, the faster the merry-go-round spins, the harder each person who stands up seems to be thrown against the wall.

After all the people on the merry-go-round have been standing for a little while, the heavier ones will move toward the wall faster than the lighter ones. If the merry-go-round is stopped suddenly, there will be more heavier people near the outer wall than lighter ones. Also, there will be more lighter people near the center than near the outer wall. The faster the merry-go-round had been spinning, the better this separation on basis of weight should be.

A merry-go-round is much the same as a spinning metal disc or pie tin. So consider a large spinning pie tin. Imagine that we have a mixture of some finely ground sawdust and finely ground sand. The sawdust and sand are so thoroughly mixed that the individual particles of each cannot be differentiated, while the grains of each are evenly distributed among one another in the mixture. How could the sand grains be separated from the wood grains of the sawdust? The spinning pie tin can be used to do so. Again, this is possible because the sand grains are heavier than the wood grains. One can simply place the sand and wood dust mixture on the spinning pie tin. Say that the mixture is spread evenly over the pie tin while it is not spinning. Also imagine that the pie tin has a large rim at its outer edge that prevents the mixture from flying off as it spins. Now say the pie tin is spun at a high rate of speed. Then the sand grains, being heavier, will be thrown harder toward the rim of the pie tin than the wood grains, which are lighter. Thus the heavier sand grains will tend to concentrate near the rim, while the lighter wood grains will tend to concentrate near the center of the pie tin. It will not be a perfect separation; some wood grains will be found among the sand grains near the rim of the disc, and some sand grains will be found near the center and middle of the disc. The mixture near the edge will be richer in sand grains, while that at the middle and near the center will be richer in wood grains.

After the sand-wood dust mixture has been spun once on the pie tin, we get two mixtures, one of which is richer in sand and the other in wood dust. We can collect each. Then we can spread the one richer-in-sand grains over the pie tin again and spin it. In this way, it is again separated into two parts, or fractions. The one near the rim will be yet richer in sand, while that at the middle and near the center will be yet richer in wood dust. We can do the same with the original fraction richer in wood dust. Then we would get a smaller fraction of mixture rich in sand grains near the rim of the pie tin, and one very rich in wood dust in the center. The process can be repeated until all the sand grains are separated from the wood grains, for all practical purposes.

The principle of the merry-go-round and the spinning pie tin in separating objects of different sizes and weights is the same as that of the centrifuge. The device is often called the ultracentrifuge because the tube holding the mixture of organelles is spun very fast—often at many thousands of revolutions per second. These devices come in many different designs. Only basic principles can be described here. But the invention of the centrifuge was a needed break for cytology. The centrifuge made the separation of different cell organelles possible.

Now let's go back to the mixture of cell organelles in the sugar solution. These organelles have different sizes and weights, like the sand and sawdust grains on the pie tin or the people on the merry-go-round. The main part of the centrifuge is a glass tube into which the mixture of organelles is placed. This tube is then spun around (FIG. 7-3) at a great speed, which can be changed at will. When a sugar solution containing broken-up cells is put into the tube of a centrifuge that is being spun at great speed, the heaviest organelles feel the greatest force that pushes them to the end of the tube, and they tend to settle near the end of the tube. This does not mean that all the heaviest organelles go to the end of the tube on the first spinning of the mixture. All organelles feel a force tending to push them out, like a man on the spinning merry-go-round feels a force tending to push him off. So all the organelles tend to go toward the end of the tube. But the heavier ones have a greater tendency to do so than the lighter ones, and feel a greater force than the lighter ones. But there are already some of the lighter organelles at the end (or bottom) of the tube before the mixture is spun; they stay there, and cannot move up the tube.

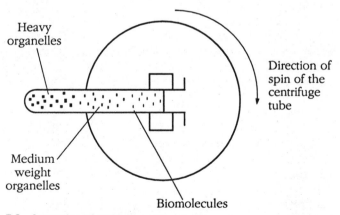

7-3 Separation of organelles with a centrifuge.

The heaviest organelles are nuclei in general. Also the disrupted cell membranes are just about as heavy. So if the mixture at the end of the tube is separated from the rest after a spinning of the tube, it will be much richer in nuclei than the rest of the liquid in the tube. How the liquid is removed from the tube need not be of concern here. There are means of doing this, although the specific designs of the many types of centrifuges in use is beyond the scope of this account. The remainder of the liquid in the tube will thus contain most organelles lighter than the cell nuclei, and if this remaining portion is again spun for a certain time in the centrifuge at a greater speed than before, the next heavier organelles, the mitochondria, go toward the end of the tube. So in this case, the liquid at the end of the tube becomes richer in mitochondria than the rest of the liquid further up the tube. If this portion is separated off, the liquid left over will be rich in organelles lighter than the mitochondria. These include the ribosoms among others. If the tube is spun at yet a higher rate, these, too, move toward the end of the tube.

What is therefore accomplished with the centrifuge is a separation of the original uniform mixture of cell organelles into parts or fractions rich in specific organelles. This was a big step forward for cytology and biochemistry. The chemistry of different organelles could now be studied more directly than was possible prior to 1923. A mixture of different intact cell organelles could be separated, or fractionated, into component mixtures richer in a given organelle. The process is sometimes called *fractionization*. The original mixture is separated into several fractions; each is richer in a given organelle than the original mixture.

Each such fraction of the original mixture is only richer in a given kind of organelle. After a first run of spins in a centrifuge, each fraction contains some organelles lighter, and some heavier, than the particular one it is richest in. However, each of these fractions of the original mixture is spun in the centrifuge tube once again, like the sand-sawdust mixture in the pie tin. In this way, they are separated into further portions or fractions which are richer in a given organelle than the original fractions. By these means, nearly pure solutions of each organelle can be obtained after a series of applications of the ultracentrifuge to the original mixture of organelles. But no fraction is really 100 percent pure in one organelle. Nevertheless, the degree of purity that could be achieved in this way with the ultracentrifuge was almost perfect, and organelles could then be analyzed chemically.

Centrifuges of many designs, efficiencies, and sizes have been put to work since the invention of the device in 1923. The centrifuge made the isolation of various organelles in large numbers possible. These could then be directly analyzed chemically, and many aspects of their composition were then no longer mysteries. With the aid of the ultracentrifuge many cell nuclei from different organisms were chemically studied. These studies confirmed that the cell nucleus generally contains a lot of nucleic acids—both DNA and RNA—although structural proteins, enzymes, and smaller amounts of elements such as calcium, magnesium, iron, and zinc where found also. It was found that the amount of DNA in cell nuclei of a given organism is constant, although the amounts of RNA could vary with different circumstances.

When almost pure samples of mitochondria were examined chemically, they were found to contain proteins in the form of enzymes that speed up the oxidation of fats, carbohydrates, and starches, as well as the oxidation of other biochemicals. These findings helped show that these organelles were indeed the power houses of the cell.

Up to 1923, biochemists and cytologists had developed four chief tools to aid them in their research: the microscope, normal chemical analysis, the use of enzyme solutions to break large biomolecules into smaller ones that comprised them, and the semipermeable membrane and other filters. They were also blessed from time to time with certain fortunate incidents that made the separation of specific organelles and proteins possible. The development of the ultracentrifuge was an addition to these already existing methods. The separation of different cell organelles was one of its accomplishments. Now almost pure samples of a given organelle could be prepared. That made it possible for a given organelle to be more easily scrutinized under the microscope, while the biochemist could then apply various chemical solutions to the organelle collection and observe their effects on the organelle with the microscope. Say a pepsin solution is applied to such a collection of a given organelle. Pepsin dissolves proteins. The organelle could then be observed under the microscope after the pepsin treatment. If it was then seen that certain structures of the organelle has been removed by the pepsin treatment, these parts of the organelle has to be protein in composition. Alcohol dissolves fat. Say now that a collection of the same kind of organelle is treated with alcohol. If the microscope reveals that other parts of the organelle have been removed by the alcohol, these parts had to be fatty in nature.

The separation of organelles was only one service the ultracentrifuge performed for our understanding of the cell and its makeup. What was the nature of the liquid portion left after most organelles had been removed after repeated spinning in the centrifuge? This liquid was found to contain many different proteins and other biochemical substances in solution. What if this fraction is spun further in an ultracentrifuge at yet higher speeds than were used to separate the organelles out? When this was done, it was found that the different large molecules in the liquid moved down the tube different distances, as the organelles had, depending on their size and weight, and this made the separation of additional proteins possible. It also made a determination of the masses of different biomolecules possible. Like the heaviest organelles had done, the largest and heaviest molecules moved to the end of the tube when it was spun rapidly, for the most part. So, again, the liquid could be broken into a number of fractions, each more pure than the original mixture in a given kind of molecule, that could be analyzed chemically.

By the mid 1930s, additional proteins had been isolated through the use of the available ultracentrifuges. The development of these devices was a great asset to cytology and molecular biology, and is a line of experimental technology that has been continually improved right up to the present time.

X-RAY STUDIES OF PROTEIN

In the early decades of the present century, another line of attack to determine the properties of biomolecules like proteins was being pursued: the study of X-ray scattering from crystals of substances. Only a brief outline can be given of this here.

X-rays are a form of electromagnetic radiation like visible light. These radiations are known to have a wave aspect and X-rays are a wave phenomenon, so let's briefly look at the behavior of simpler, more familiar waves like water waves to get some idea of what is involved in X-ray scattering by crystals.

Consider a lake or pond with a series of water waves moving over its surface. It is clear that the waves move over the water surface. But look at the waves more closely. We see that the water surface in the area of the moving waves is made up of a series of moving elevations and depressions of the water surface. Each elevation is preceded by a depression which, in turn, is preceded by another elevation. Each elevation in the wave train has a depression both in front and in back of it. While each depression has an elevation both in front and in back of it. These alternate elevations and depressions move over the water surface together. The moving elevations and depressions of the water surface may be caused by a moving object floating on the surface, a wind that blows against the water surface, or by a stone thrown into the water, for example. But the important point to note is that each moving wave is made up of a moving elevation and depression of the water surface. The elevation is known as the *crest* of the wave. The depression is called the *trough* of the wave. Each wave therefore consists of a moving crest and a trough.

Now consider two adjacent water waves in the wave train. The distance from a given point on the crest of one of the waves to the corresponding point on the crest of the next, or from a given point on the trough of one wave to the corresponding point on the trough of the other wave, is the *wavelength* of the wave motion. The number of waves that pass a given point on the water surface each second is known as the *frequency* of the wave motion. Water waves can serve as a model for all wave phenomena. Wavelength and frequency are properties of all types of wave motion. The wavelength of water waves depends on the strength of the disturbance that produces them.

Visible light and other electromagnetic radiations have been shown to be a wave phenomenon. In this case, the waves are made up of changing electric and magnetic forces that move through empty space. Electromagnetic waves come in a broad spectrum of wavelengths and frequencies.

To briefly talk about these, it will be helpful to mention the system of length measurement used by scientists, the *metric system*. The basic unit of length in that system is the *meter*. One meter (represented by m) equals 39.37 inches. One *centimeter* equals one hundredth of a meter, or .01 m. The centimeter is represented by cm. A *millimeter* is one thousandth of a meter, or .001 m; the millimeter is represented by mm. One centimeter, or 1 cm, equals 2.54 inches. When physicists and chemists began to talk about the sizes of small objects like atoms and molecules, they had to define more convenient units of length than these. Since 100 million hydrogen atoms laid end to end would form a line only 1 inch or 2.54 cm in length, it would be quite inconvenient to express the diameter of an atom in meters or centimeters. We shall see shortly that the same consideration applies to the wavelengths of many common electromagnetic radiations. So a more convenient unit of measurement is necessary to deal with these situations. One such unit is the *nanometer*, represented by nm, and one nanometer equals .000000001 m. All these units are convenient in our discussion of electromagnetic radiations and other biochemical and cytological objects.

Electromagnetic radiation comes in a wide spectrum of wavelengths. The longest of these belong to radio waves. Radio waves have wavelengths ranging from about 100,000 m to 10 cm. They are followed by microwaves, which range in wavelength from 10 cm to 1 mm. Then there are extremely short-wave microwaves that have wavelengths in the sub-millimeter range. Then comes infrared radiation. Generally this radiation has wavelengths much smaller than one millimeter. It is at this point that the nanometer becomes more convenient to use. Infrared radiation ranges in wavelength from about a little less than 1,000,000 nm to about 700nm. Then comes visible light, from 700 to 400 nm. The color of light depends on its wavelength. The longest wavelengths of light, ranging from 700 nm to 650 nm, look red. The range from 650 nm to 600 nm belong to orange light, while yellow light belongs to a wavelength span from 600 nm to about 560 nm. From 560 nm to 500 nm we have green light, while from 500 nm to 440 nm we have blue light. The shortest wavelengths of visible light belong to violet light: these range from 440 nm to about 400 nm. Although the waves of visible light are small, they are several hundred times bigger than the average atom in size.

The relationship between the size of the object being probed with a given kind of electromagnetic radiation and the wavelength of the radiation is important when it comes to microscopic investigations and X-ray scattering.

Imagine that a large piece of wood floats on the water surface as the wave train passes by. And say that alongside the piece of wood a small piece of cork floats. As the water waves come upon the piece of wood, they will be reflected by it in certain directions, though they will hardly be disturbed by the small piece of cork. Why? It all has to do with the size of the wavelength of the wave motion.

In this example the dimensions of the large wood block are much larger than the wavelength of the water waves, and therefore, the waves are significantly disturbed or reflected by it. The cork, however, is very small compared to the wavelength of the waves, and thus has very little affect on them.

An expert in the physics of waves would be able to deduce various features of the shape of the wood block by studying the water waves reflected from it. He could do the same for a large ship or boat on the water that scatters these waves.

This relationship between size and wavelength is also important in the limitations of the light microscope in cytology and biology in general. We see objects around us because they reflect light waves. They do this well because the wavelength of light is so small compared to their dimensions. Most cells are large enough to reflect light waves, so we can see them under the light microscope when they are properly stained, while the same is true for some cell organelles. But many organelles are about the same size or smaller than the wavelengths of visible light. Thus many of them could not be seen clearly, even with the newly improved light microscopes of the 1930s. This presented a problem for biological science of the time. Some of the alternatives to visible light used to get around this problem in the field of microscopy will be examined in the next chapter.

Most large molecules and the atoms comprising them are much smaller than the waves of visible light, and therefore cannot be seen with a light microscope, no matter how how its magnification. But by 1912 a new kind of electromagnetic radiation had been discovered that had a much shorter wavelength than those of visible light. It had become known as X-rays. The wavelengths of X-rays range from 1 nm to 0.01 nm, the same as the range of sizes of atoms and molecules. Around the beginning of the century, it occurred to some physicists that X-rays could be used to probe the arrangements of atoms in crystals of substances in much the same way that water waves and light waves can be used to give information about the properties of larger bodies. Experimenters directed beams of X-rays at crystals of various substances and studied the way the X-rays were scattered by the molecules in the crystals. Through a mathematical analysis of the manner in which the X-rays were scattered by the crystals, the arrangements of the atoms and molecules in the crystals could be determined. This field of study came to be known as X-ray crystallography. It all started in 1912.

In that year the arrangement of the sodium and chlorine particles in the cube-shaped crystals of table salt had been determined by X-ray scattering. The crystalline structure of other inorganic substances were figured out in the 1920s by X-ray scattering. These substances had simple molecules. They were thus easier to study by X-ray methods. Two pioneers in the field of X-ray crystallography were Lawrence Bragg, an English physicist who started determining the basic principles of the new field in 1912, and the great American chemist Linus Pauling, both of whom had figured out the basic mathematical techniques of X-ray analysis in 1930.

But the X-ray scattering methods that worked with simpler substances ran into considerable difficulties when applied to organic ones like proteins and nucleic acids in the early 1930s. These substances had more complex molecules. Before turning to these problems, let's say a few words about how X-ray studies are carried out.

In simplified terms, a beam of X-rays from an X-ray tube is sent through a group of crystals of the substance being studied. After passing through the crystals, the scattered X-rays fall on a photographic plate. A photographic plate is simply a sheet of film containing an emulsion that darkens at the places where X-rays strike it after it is chemically developed. Thus after the plate is developed, a pattern of dark spots appears on it. The darkened regions show where the scattered X-rays in the beams hit the plate. The pattern and arrangement of the dark dots on the plate can be analyzed mathematically. When they are, they give information about the spatial arrangement of atoms in the crystals of the substance. The pattern of dark dots tells the expert X-ray crystallographer the way the X-ray waves were scattered by the atoms. That, in turn, tells him how the atoms are arranged. It takes much training and experience to interpret X-ray scattering patterns.

In the 1920s, X-ray patterns of simple substances could be easily obtained. These substances had simple molecules. Such simple particles group up in a neat crystalline pattern when the crystals are dry. But, as we have seen, molecules of proteins and nucleic acids are made up of long chains of atomic groups. Such complex chains of atoms can assume many different arrangements with respect to each other in dry crystals of the substance, and X-ray analysis of such crystals was first attempted in the 1930s. But because of the helter-skelter arrangement of the atoms in dry protein crystals, good scattering patterns were not achieved, although some order in the patterns could be seen. That gave some encouragement for further studies of the same kind.

Crystals of specific proteins were often obtained through precipitation out of solutions from cells, and these crystals contained a lot of water mixed with the protein. Much of this water remained in the solid crystals for some time. Such crystals were thus wet or moist. Small water molecules were dispersed between the large protein molecules in the crystals. These kept the larger molecules from being too compact in arrangement. They were then freer to orient themselves in a more orderly way to give a more perfect crystal. The large protein molecules could then line up with one another in the crystal to give a clearer, more ordered X-ray scattering pattern. By the 1930s, biochemists learned how to grow certain protein crystals in solution so that much better X-ray scattering patterns of proteins could be obtained, though the patterns for such large molecules were not easy to interpret then.

In 1937, the noted chemist, Linus Pauling, and another expert, Roger Corey, began to examine the X-ray patterns of the simpler straight-chain proteins as well as those of simple amino acids. By the 1950s, this work had contributed much to our understanding of protein structure.

But X-ray studies, though they shed much light on protein and nucleic acid structure, could not alone do much in the absence of other methods. X-ray crystallography, like other biochemical techniques, had to be combined with other modes of investigation, some of which have been discussed and some of which have not. Pauling's and Corey's research with X-rays was proceeding at the same time as many other lines of protein and nucleic acid research, and went on through the 1940s and into the 1950s. It gave important clues to the problem of protein structure in itself.

To really be effective, X-ray scattering had to be used in conjunction with pure biochemical studies. But through the 1930s such studies were hampered by the complexity of protein solutions gotten from living organisms, and no systematic method to separate individual proteins out of such solutions had been developed. The problem became more involved when a protein was broken down into its amino acids with an acid or pepsin solution. How could the many amino acids in such a mixture be separated from each other and analyzed for their amounts? This had to be done before the primary structure of even simple peptide chain proteins could be determined. The primary structure of protein molecules was the first problem that demanded the attention of experts in the 1930s. The primary structure of a protein is made up of four items. Each had to be determined. The first of them is the number of peptide chains in the protein molecule. The second is the particular amino acid units in each of the peptide chains. Then there is the number of each kind of amino acid unit in each of the chains, while the fourth aspect is the arrangement of the amino acid units in each chain. To determine these four factors seemed to be an imposing problem when Pauling and Corey started their X-ray studies in 1937.

That was why there was so much interest in X-ray studies of proteins in the 1930s. A direct chemical attack on the problem seemed to be out of the question because of the inability to separate the components of the mixtures involved, so that X-ray studies seemed to be the only hope for determining the structure of the few proteins that had been isolated at the time. But it became apparent that the solution of the problem by that route was going to take a long time. Fortunately, another biochemical technique had come to light in the 1930s that, by the 1940s, had shed much light not only on the problems of the separation and structure of different proteins, but also on that of nucleic acids. This new method led up to the Watson-Crick model of DNA structure more than any other method of the time had done. Of course, other methods helped solve the DNA problem, but this new method made the difference. Let's look at the method.

Chapter **8**

Later twentieth century methods

The components of protein mixtures were very difficult to separate from each other in the 1930s. To gain some insight into the problem, let's look at some ordinary methods used by chemists to separate the components of the mixtures they encounter.

DISTILLATION

First consider some simple examples. Take a mixture of sugar dissolved in water. The mixture is uniform and there at first seems to be no simple way to separate the sugar from the water. But chemists do so easily. They make use of the simple fact that water evaporates much more easily than sugar. That is, water is a liquid that boils at a relatively low temperature—100 degrees celsius—while sugar is a solid that shows no tendency to evaporate. This says that water is more volatile than sugar. To separate the components of this mixture, chemists simply boil the mixture and condense the vapors given off, and these vapors prove to be water after they are condensed. The sugar is left behind as a solid as the water of the mixture boils off. The property that makes this separation possible is a difference in boiling point between sugar and water. Sugar decomposes into carbon and water before it boils. However, if it could boil, all evidence indicates that its boiling point would be higher than that of water.

The two components of a mixture of table salt dissolved in water can be separated in exactly the same way. The water can be boiled off and condensed, leaving the salt behind.

Now consider a mixture of roughly equal amounts of two liquids— say alcohol and water. In this case the alcohol boils at a lower temperature than the water, which again means that alcohol is more volatile than water. In the same way as before, chemists can separate the alcohol from the water by boiling the mixture and condensing the vapors given off. By boiling the mixture for a while and condensing the vapors, the chemist can separate the original mixture into two fractions, and the fraction gathered from the condensed vapors given off will be found to be richer in alcohol, while the fraction left behind will be found to be richer in water. That is because alcohol is more volatile than water. Thus more alcohol boils off the mixture than water. Both of the resulting fractions can be boiled again to give additional fractions richer in both components of the mixture, and eventually all the water can be separated from the alcohol.

The same technique works if we have a mixture of several liquids that all boil at different temperatures. When such a liquid mixture is boiled, the components having the lower boiling points tend to boil off first. Thus the fraction of liquid obtained by condensing the vapors given off right after the mixture starts to boil is richer in these components than in the ones having the highest boiling temperatures. Again the resulting fractions can be boiled. This can be repeated until the several liquids are separated from each other. The more volatile liquids boil off first.

The process of separating the components of a mixture by taking advantage of the difference in volatility of the components through boiling is called *distillation*. From these examples of distillation, we can see one example of the chief situation the chemist takes advantage of in separating mixtures into their component substances. This is a difference in characteristics or properties of the components of the mixture. In these examples, it is a difference in boiling points, or in volatility, which is the ease of evaporation of the components. There are many other kinds of differences in properties that aid chemists in these ways.

Consider a mixture of finely ground sand and salt. Imagine that the salt and sand grains are so small and so thoroughly mixed that they cannot be separated by mechanical means. Then how can the sand and salt be separated from each other? Salt dissolves in water, while sand does not dissolve in water, or is insoluble in water. Advantage can be taken of this difference in properties of salt and water to separate them. Simply add an adequate amount of water to the solid mixture and stir the resulting mixture. The salt will dissolve in the water, leaving the sand behind to sink to the bottom of the mixture. Now we have a salt solution with sand at the bottom. But how can the salt solution be separated from the sand? The overall mixture of salt, sand, and water can be poured into a funnel fitted with a sheet of filter paper that is held over a beaker or flask. The liquid portion of the mixture easily passes through the filter paper. The sand grains do not. So the salt solution flows down the spout of the funnel into the beaker or flask. The wet sand left in the funnel can be rinsed with additional water to wash any remaining salt solution into the beaker or flask. The filtering process has thus separated the sand from the salt solution, and the salt can now be recovered from the salt solution by distillation. The sand has been separated from the salt through a difference in solubility between sand and salt in water.

Imagine a mixture of finely ground sand and wood grains. How can the sand and wood grains be separated from each other? Sand is denser than water, while wood is less dense. Bodies denser than water sink when placed in it. Those that are less dense than water float on water when placed in it. If water is added to the mixture of sand and wood grains, the sand grains sink to the bottom of the water, while the wood grains float to the top. Then, the top portion of the resulting mixture can be separated from that at the bottom, and both portions can be filtered to recover the wood and sand grains after they are dried.

DIFFUSION

There are also means of separating the components of mixture of gases, one of which involves the process of *diffusion*. Some solid materials contain small, submicroscopic holes or pores in them that are of the same order of size as atoms and molecules. Because of these pores, gases can slowly move or diffuse through a sheet of such a porous material. Now imagine a mixture of oxygen and hydrogen that is put into chamber A of the imaginary device shown FIG. 8-1. At the start of this experiment, assume that chamber B consists of a vacuum. (In practice, a perfect vacuum is not attainable, and in reality B would be a nearly perfect vacuum.) It would be found that the gaseous mixture in chamber A would slowly move or diffuse through the sheet of porous material C into chamber B. But it would be found that the lighter gas, hydrogen, would diffuse into B through the sheet faster than the heavier one, oxygen. The lighter a gas at

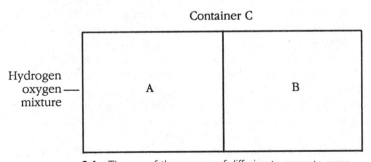

8-1 The use of the process of diffusion to separate gases.

a given temperature and pressure, the faster it diffuses through a porous sheet of material. Say the original mixture in A contains an equal number of molecules of oxygen and hydrogen; that is, it contains equal numbers of moles of the two gases. Hydrogen diffuses into B faster than oxygen. So, after a certain time, the mixture in B will be richer in hydrogen than in oxygen. The remaining portion in A will then be richer in oxygen than in hydrogen. The two resulting fractions of the original mixture can be put through the same diffusion process many times until the hydrogen and oxygen are separated from each other.

In this case, the basis of the separation is the difference in rates at which the two gases diffuse through a porous sheet or membrane. There is a physical explanation for this. At a given temperature, lighter molecules move faster than heavier ones. The oxygen molecules are each 16 times as heavy as the hydrogen molecule, and, at a given temperature, the hydrogen molecules in the mixture, on the average, move four times as fast as those of oxygen. Therefore, they move through the pores in the sheet four times as fast as the oxygen molecules.

In all these examples one fact stands out: the means used to separate the components of a mixture depend on the difference in properties of the components. But the difference in properties of the substances depends on the difference between their molecular structures. Different substances have different properties because of the differences in the way their molecules are constructed. The two substances in the sugar-water mixture are sugar and water, and sugar molecules are quite different from those of water. They are much larger, for one thing, and have the molecular formula $C_{12}H_{22}O_{11}$. Water molecules are much smaller and have the molecular formula H_2O. Because of this, in part, the cohesive forces between the larger sugar molecules are stronger than those between the water molecules. Therefore sugar is a solid while water is a liquid at the same temperature. This is also the reason why sugar is much less volatile than water. The oxygen molecule is 16 times as heavy as the hydrogen molecule. Thus, it moves much slower at the same temperature and diffuses more slowly than the faster moving hydrogen molecules through a porous membrane. Big differences in properties between two substances such as these exist because of a big difference in their molecular structure.

This also means that the more similar the molecular structure of substances are, the less the measurable properties of the substances will differ. So, generally, mixtures of such substances should be harder to separate into their components. This has often proved to be true. But there is some difference in molecular structure between any two substances. So there must also be some differences in their properties. However, since these differences are not great in this case, more sensitive methods must be developed to separate the substances.

This was the situation facing chemists in the early decades of the century when it came to separating the components of mixtures of proteins and nucleic acids. Peptide chains are generally similar in molecular structure. The same can be said for amino acids. That is why they were difficult to separate from mixtures. If a biochemist was fortunate, his methods could sometimes get a certain protein to crystallize out of solution, though that was the extent of it.

CHROMATOGRAPHY

That was the situation that faced a Russian botanist Mikhail Semenovich Tswett in 1906 with regard to a complex mixture of plant pigments he wished to separate from each other. But none of the chemical techniques available to him could accomplish this, since the properties of the pigments in this mixture were too similar to lend to any kind of easy separation of the pigments. Tswett made another chance discovery that was to revolutionize protein and nucleic acid chemistry four decades later. He happened to have at his disposal some of the material alumina, or aluminum oxide Al_2O_3, in powdered form. So he filled a tube with powdered alumina, so that it formed a long column, and somehow he decided to pour the complex mixture of plant pigments down the column of powdered alumina in the tube. When he did so, he discovered a peculiar phenomenon. As the mixture seeped down the column of powdered alumina, it separated into a series of bands along the column, each of which had a different color. Tswett could determine that each colored portion (or band) on the column of alumina was composed primarily of one of the plant pigments that made up the organic mixture. He could separate the components of the liquid mixture by letting the mixture move down a column of powdered alumina. This was the first example of what is today called column *chromatography*, and of chromatography in general. This technique of separating the substances in a complex mixture was called chromatography because it was first used to separate colored substances.

Not much attention was paid to Tswett's discovery in the first two decades of the twentieth century. But that changed in the 1920s. By the 1930s the technique of column chromatography could be used to separate the proteins in a mixture when one could obtain a sufficient amount of the mixture.

Why does the procedure work? What is its physical basis? Consider the mixture of plant pigments again. The mixture consists of many different pigments dissolved in some organic liquid, and is composed of many different pigment molecules. The powdered alumina consists of small grains of alumina. When the mixture is poured down the column, the liquid flows through the spaces between these grains, and the different pigment molecules in the liquid are attracted to the grains to different degrees. Some molecules are strongly attracted to them. Others are weakly attracted to them. Some molecules are moderately attracted to them. But each kind of pigment molecule is attracted to them to a different degree. Now the degree of attraction between a given kind of pigment molecule and the grains determines how strongly that pigment adheres to the alumina as it moves down the column. The stronger the particular pigment adheres to the grains, the slower it moves down the column. The weaker a pigment is bound to the grains, the faster it moves down the column of alumina. Each pigment should migrate down the column at a different rate. Thus, they separate from one another as they move down the column.

The same applies to other liquid mixtures of proteins that can be obtained in significant quantity. Each protein in the mixture moves down the column at its own rate that is different from that of any other protein in the mixture. This technique of chromatography is known as *column chromatography*. It has made the separation of proteins from complex mixtures possible. The particular material that fills the column in this kind of chromatography is known as the *absorbant*. Alumina is only one absorbant. Some others are Charcoal (or carbon), lime-CaO, calcium carbonate (CaCO), and starch. But the process has one important limitation now as well as in the 1930s: one must have a significant amount (usually a large amount) of the mixture if the technique is to be useful. The trouble was that enzymes and other important proteins could only be gotten from cells in very small amounts. Extracts from cells contained only minute amounts of various enzymes. Such small amounts of substances would be very difficult to isolate by column chromatography. And when it came to analyzing these proteins for their amino acid content, the situation was even more unworkable, so that column chromatography did not really help biochemists when it came to separating mixtures of enzymes and nucleic acids.

But another fortunate break came in the 1940s. In 1944, two biochemists, Archer Martin and Richard Sygne of England, developed a new kind of chromatography. The new method was well suited to analyzing mixtures holding only small amounts of different proteins.

The method utilizes a piece of filter paper as an absorbant. Often a small amount of the mixture to be analyzed is placed on a certain portion of the paper and is allowed to dry. That is, the liquid in which the proteins are dissolved evaporates from the paper, leaving the proteins in the mixture on the paper. Then the edge of the piece of paper is dipped in a liquid in which all the components of the mixture are soluable. Such a liquid—a *solvent* as it is called—can be found by experimenting on samples of the dried mixture. The liquid then seeps up the paper through *capillary action*. Capillary action is simply the rise of the liquid solvent up the paper seemingly by itself. It is caused by the fact that the molecules of the solvent are attracted to those in the fibers of the paper, and are pulled up the paper as a group by this attraction—a type of attraction between different molecules known as *adhesion*. So the liquid solvent moves up the paper until it reaches the portion of the paper on which the mixture of small amounts of the proteins or amino acids is situated. These substances then dissolve in the liquid layer as it passes them. Now we have a state of affairs similar to that which prevailed in column chromatography. The different substances in the mixture are carried along the paper by the solvent as it moves up the paper. But the various protein molecules among those of the moving solvent are attracted to the fibers of the paper to different degrees. Again those that are most strongly attracted to the fibers move up the paper with the solvent at the slowest rate, while those that are most weakly attracted to the fibers move the quickest with the solvent. So again each substance of the mixture moves up the paper at its own rate. Thus, as the protein mixture moves up the filter paper, the different proteins are separated from each other to a great degree. But the separation is not perfect. When the paper is dried by evaporation of the solvent, the original mixture has been separated into a number of bands on the paper (FIG. 8-2). These bands can be made visible by various chemical tests to be described later. Each band is particularly rich in one particular component of the original mixture, though it is not pure in that component. The separation of the substances is not yet complete.

A Bands on filter paper

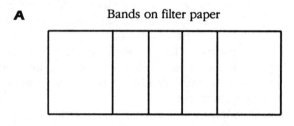

B Spots of amino acids after paper is rotated

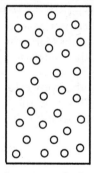

8-2 The use of paper chromatography for analyzing mixtures.

The bands on the paper are richer in one component of the mixture than the original mixture had been. So the paper is turned around through a 90 degree angle, and its end is again immersed in the solvent. The solvent again seeps up the paper again through capillary action until it reaches the series of bands of proteins, and as it moves by each band, the different proteins in each are separated from each other further. The result is that each band is broken up into a series of parts or blobs on the paper, each of which is nearly pure in one particular component of the mixture.

This technique of chromatography is known as *paper chromatography*. It proved much more able to deal with mixtures containing small amounts of proteins and amino acids, because it dealt with analyzing smaller amounts of the mixture. The small blobs of different substances obtained on the dried paper could be made visible by certain chemical techniques, and then analyzed. The small amount of the substances composing them would have gotten lost in a large column of absorbent and would thus be impossible to detect. But they gather at one spot on a small piece of filter paper.

Paper chromatography had another important use. Not only could it be put to use to separate various proteins, amino acids, and nucleic acids from each other, but could also be used to figure out the amino acid content of a given protein. Once a small amount of the given protein had been obtained, it could be broken down into its component amino acids through acid or pepsin treatment among others, and the resulting mixture of amino acids could then be placed on a piece of filter paper and dried. Then a solvent that dissolves all the amino acids is applied to the paper. In this way, a series of blobs form on the paper. Each is made of a given amino acid for the most part. How these are made visible and analyzed for the amounts of the amino acids they contain will be touched upon later. A given nucleic acid could also be analyzed for its nucleotide content with paper chromatography.

Paper chromatography was a gigantic step forward in the mid 1940s. Until its advent, biochemists were at a loss not only to determine the composition of specific proteins and nucleic acids coming in small amounts, but also could hardly do anything with mixtures of them. Paper chromatography was a big step on the road leading to modern genetics. It is no accident that the Watson-Crick model of the DNA molecule was derived less than ten years after Martin and Sygne introduced this kind of chromatography. More of the details will be seen in the following chapters.

In the late 1940s, another powerful technique of investigation had come into prominence. This was the method of *radioactive tracing*. Along with paper chromatography, it greatly aided biochemical and genetic research.

ELECTRIC CHARGE

The Hershey-Chase experiment was one example of historic biochemical research that made use of radioactive isotopes. To understand what these are, and to shed light on discussions to follow, let's review what modern science had learned about atomic structure by the late 1940s.

Atoms, the building blocks of the chemical elements, were thought to be solid and unchangeable during the latter half of the nineteenth century. After all, that is how Dalton, the founder of modern atomic theory, had defined them. But by the end of that century, this model of the atom left a lot of important questions unanswered. Why are atoms of different elements different from each other? What holds atoms together in molecules? Why could one oxygen atom unite with only two hydrogen atoms to form a water molecule, and no more? Why does one nitrogen atom unite with three hydrogen atoms to form the ammonia molecule, and no more? Why should a hydrogen molecule have only two hydrogen atoms instead of two, three, four, or any number? In general, why were only certain atomic combinations in molecules possible while others never occurred? Nineteenth-century chemistry, based on the solid unbreakable atom of Dalton, had no answers to these questions.

This view of the atom began to change in the last decade of the nineteenth century. At that time physicists began to carry out electrical investigations involving gases.

Electricity is an elusive but familiar phenomenon. It has been noted that if two glass rods were rubbed with silk cloth, they would repel each other when brought close together, as would two rubber rods so rubbed. But if a rubber rod rubbed with silk cloth was brought near a glass one that had also been rubbed with silk cloth, the two rods attracted each other. The attraction and repulsion displayed in these simple experiments is one example of the *electric force*. The glass and rubber rod after being rubbed with silk were said to be *electrified*, or they carried an *electric charge*. It was found that many other bodies could be electrified in the same way. But any given electrified body either repelled a glass rod after it was rubbed with silk or attracted it, and also either attracted or repelled a rubber rod so rubbed. If it attracted the glass rod, it repelled the rubber one. If it repelled the glass rod, it attracted the rubber one. In no case did it repel both or attract both. These findings suggested that there were two kinds of electricity. The charge on an electrified glass rod came to be called *positive* charge. That on an electrified rubber rod came to be called *negative* charge. The above findings indicated that like charges repel and unlike charges attract. Two electrified glass rods or two electrified rubber ones are likely charged—positively charged in the first case and negatively charged in the second. Thus two such rods repel each other. An electrified glass rod attracts an electrified rubber one, since these objects are oppositely charged. The harder the two rods were rubbed with silk, the stronger the electrical attraction between them became. This indicated that electric charge was a quantity that came in different amounts and could be measured.

The electric charge on the glass and rubber rods is an example of *static electricity*. The charge just stands still on the objects, so to speak. But it was also found that electric charge could not move easily through these materials. For this reason, they held on to their charge. It could not leak off them to surrounding bodies. Metals were found to be different in this way. An electrified metal object seemed to lose its charge easily. Also if two oppositely charged, electrified, metal objects were connected by a metal wire, the electric charge would move through the wire until each became electrically neutral. The charge flowing through the wire is charge in motion; it is an example of an *electric current*, or moving electric charge. Some sources of electric current are batteries and electric generators.

In the last decades of the nineteenth century, experiments were being done that involved passing electric currents through gases such as hydrogen, oxygen, nitrogen, air, and others. In 1897 the great English physicist J. J. Thompson was studying a peculiar phenomenon in this area. It occurred when an electric current was passed through a gas of very low pressure in a glass tube. The tube used for these studies is shown in FIG. 8-3.

Tube

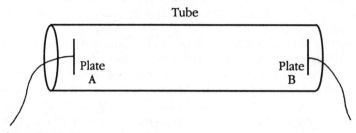

8-3 A gas tube used in the study of electrons.

Inside the tube were two metal plates A and B, one at each end of the tube, that were connected to the outside of the tube by wires that could be attached to a variable electric power source. The wires were connected to the power source so that A became negatively charged and B positively charged. The plate A is called the *cathode* and plate B the *anode*. If a lot of a gas like air was in the tube, the charge would just build up on the plates, and the most that would occur is that sparks would jump from the cathode to the anode.

But something much more interesting occurred with only a very small amount of gas in the tube. Then, when the plates were charged, it appeared that greenish rays would emanate from the cathode and go toward the anode. They caused a greenish glow at the anode's end of the tube when they struck it. These rays became known as *cathode rays* because they seemed to emanate from the cathode.

Thompson made some interesting discoveries about the cathode rays and wondered about their nature. It was found that they moved in straight lines from the cathode to anode. This was shown to be the case by the fact that objects placed in their path caused a shadow in the greenish glow at the end of the tube. Also electric and magnetic forces had the effect of bending the rays out their usual straight line paths when such forces were applied. That was shown by the fact that the position of the glow at the anode end of the tube changed its position when a magnet was brought near the tube. The same thing would happen if the rays were passed between two oppositely charged plates in the tube. The position of the glow changed as more charge was put on the plates. When they were uncharged, the rays followed their normal straight line paths.

Thompson pondered the question: what was the real nature of the cathode rays? Were they made of moving charged particles, or were they an electromagnetic radiation like light, X-rays, or radio waves? The fact that they cast shadows of objects placed in their path did not really show anything decisive at the time. After all sunlight and other light—an example of electromagnetic radiation—does the same when objects are placed in its path. But no form of electromagnetic radiation was known to be affected by electric and magnetic forces. Cathode rays were. That showed that they were a beam of electrically charged particles being emitted from the cathode. The rays were attracted to the positively charged plate, and repelled by a negatively charged one, so that the cathode ray particles had to have a negative charge. By more complex experiments, Thompson was also able to determine their mass, or weight, and they proved to be much lighter than atoms. Each of them had a weight of about $1/1840$ that of the hydrogen atom, the lightest atom known. The cathode ray particles were named *electrons*.

What, really, were these particles? The only entity flowing though the wires connected to the tube and on to the plates of the tube was an electric current. However, an electric current is only electric charge in motion, while the rays were composed of negatively charged particles. This led physicists to a reasonable conclusion: the electrons of cathode rays had to be the smallest particles of negative electric charge. They also arrived at a second conclusion: negative charge is the only kind of charge that moves through matter as an electric current. This had to be true since no positively charged cathode ray particles were found in these first experiments. So, by 1900, the electron was seen as the smallest unit of negative charge possible.

But what else were they? And also, where is the positive charge in matter located? These findings had indicated that a negatively charged body had an excess of electrons, while a positively charged one had a deficiency of them. An electrically neutral body had just enough electrons to balance the positive charge it contained. At this point another fact was noted in cathode ray experiments: the nature of the electrons emitted by the cathode were the same no matter what metallic element made up the cathode. From this, Thompson concluded that electrons must be present in all kinds of matter, and he postulated that they were one of the smaller particles that composed the atom. This conclusion was strengthened by two other physical phenomena that were discovered around the same time as the cathode rays.

One of these was the photoelectric effect. It was discovered that light could cause some metals to emit electrons when it fell on them. These electrons proved to be the same as the cathode ray electrons, having the same weight and charge. They were also emitted by many metals exposed to different kinds of electromagnetic radiation, which again implied that all kinds of atoms must contain electrons.

RADIOACTIVITY

Then there were *radioactive* elements. Uranium and radium were the chief examples of these. These heavy metallic elements continually gave off a powerful radiation that did not die down with time to any noticeable extent; the radiation also darkened photographic plates in contact with the elements. What was the nature of this powerful radiation? The radiation could be absorbed only by thick sheets of lead or concrete in many cases, a fact that was used to devise an experiment to find out what the radiation was composed of. A sample of a radioactive element was placed at the bottom of a deep hole in a large lead block B, as shown in FIG. 8-4. The block absorbed all the radiation except for a narrow beam that escaped through the hole at the top of the block. Above the block was a photographic plate P. The whole arrangement was put into the field of a large magnet M in such a way that the magnetic force was directed perpendicular to the radiation beam emerging from the hole. The pattern on the photographic plate left by the radiation showed that the beam of radiation was split into three parts by the magnetic field. One part was deflected a little in one direction from that of the original beam; this component of the radiation became known as *alpha rays*. Another portion was deflected to a greater extent in the opposite direction. This portion of the radiation became known as *beta rays*. A third portion was not deflected at all. These rays were called *gamma rays*.

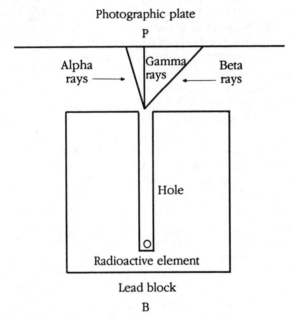

8-4 The deflection of radiation depends on its energy.

Since the gamma rays were not affected by the magnetic field of the magnet, it was quickly concluded that they were a form of electromagnetic radiation like light and X-rays. It was also found that the gamma rays had a very short wavelength—even shorter than those of X-rays—and were very penetrating, and only thick sheets of lead or concrete could absorb them. Because the alpha and beta ray components of the radioactive rays were deflected by a magnetic field, they had to be composed of moving charged particles. The alpha rays were shown to be made of positively charged particles, and were found to have a weight of about four times that of the hydrogen atom; they were the least penetrating of the three types of radiation. Thin sheets of paper could stop them. The beta rays proved to be made of moving electrons. Beta rays were found to be more penetrating than alpha rays, but much less penetrating than gamma rays. The important point is that electrons turned out to be present in the radiation from radioactive elements also.

This was enough circumstantial evidence to convince many physicists at the beginning of the twentieth century that atoms must contain electrons. J. J. Thompson then proposed a model of the structure of the atom, and envisioned the atom as a ball of positive charge through which electrons were dispersed, like raisins in a cake. In the Thompson atomic model, the total charge of the electrons equaled that on the ball of positive charge in size. This made the atom electrically neutral as a whole.

In 1911, the great English physicist Ernest Rutherford did experiments that cast much doubt on the validity of Thompson's model. In his view, the particles of alpha rays were interesting. In addition to being four times as heavy as the hydrogen atom, they were shown to have a positive charge that was twice the size of the negative charge on the electron. Rutherford was engaged in experiments that involved shooting a beam of alpha particles from a radioactive element at thin sheets of a heavy metallic element like gold. The metal sheets were very thin—only several thousand atoms thick. In these experiments, Rutherford was trying to test the validity of the Thompson atomic model. Most of the heavy alpha particles underwent only a small deflection as they passed through the metal foils. That was just what was expected according to the Thompson model. Calculations showed that the repulsion the alpha particles felt, as they passed through the positively charged ball of the atom, should be very small. Also any attraction the alpha particles experienced as they passed by the oppositely charged electrons in the ball should not have deflected them to any great extent. After all, the alpha particles were nearly eight thousand times as heavy as the electron, and therefore should not be deflected by it very much, just as a large bowling ball should not be deflected from its path of motion by a small marble in its path. So the deflection of the alpha particles by the metal foils should be small in all cases according to Thompson's model. And, most of the deflections Rutherford observed were small.

That model held true only for most of them. A small minority behaved in a very peculiar manner and were deflected through very large angles. A few bounced back from the foil toward the radioactive source that emitted them. Rutherford pondered this observation; it had to be significant. If Thompson's model was valid, all alpha particle deflections should be small. There was just no way any of them could be large. Only one solution seemed to present itself to Rutherford. The positive charge of the atom had to be concentrated in a small lump in the center of the atom, while the electrons were situated outside the lump at great distances from it—on the atomic scale of distance—and from each other. The lump of positive charge came to be known as the *atomic nucleus,* because it is placed at the center of the atom. He proposed that the negatively charged electrons moved around the nucleus, like planets about the sun, and were held in their orbits by the attraction of the oppositely charged atomic nucleus.

Rutherford's "solar system" model of the atom was only a hypothesis at the time. Yet it explained his findings well. In this model, most of the atom is made up of empty space. Therefore, the heavy alpha particles easily penetrated the mostly empty atom in the experiments with thin metal foils. The light orbiting electrons should not cause them much deflection. Also most alpha particles passed by the central nucleus at great distances as they passed through the atom. These were repelled only weakly by it. Therefore, they underwent only small deflections. But a small portion of the alpha particles passed by the nucleus at small distances, or headed right for it. These felt great repulsion, and hence, great deflection. The Rutherford model of the atom explained the results of alpha particle scattering experiments. Thompson's model could not.

So, early in the twentieth century, Rutherford's ideas about atomic structure were accepted on experimental grounds, though there were theoretical difficulties. Electromagnetic theory implied that the orbiting electrons in Rutherford's atom should constantly emit energy in the form of electromagnetic radiation of all wavelengths. Thus each electron should constantly lose energy and should fall toward the nucleus as a result. All atoms should collapse in this model according to the accepted classical theories of physics. Yet experiments confirmed the theory, and, in reality all atoms are stable. This had to be true, or the world as we know it could not exist.

Rutherford went on to make other discoveries that involved the phenomenon of radioactivity. By the early twentieth century, chemists had made accurate determinations of the relative weights of atoms. Was it just coincidence that the alpha particle was four times as heavy as the hydrogen atom? The alpha particle was about the same mass as the next heavier atom, the helium atom. Rutherford had also noted another peculiar fact. Many minerals containing the ores of radioactive elements like uranium were found to contain small amounts of helium gas, and this only occurred in minerals that held ores of radioactive elements. That was strange: finding a chemically inert gaseous element like helium in the ores of such heavy metals and not in other ores. These observations inspired Rutherford to form a curious hypothesis. What if the alpha particle of alpha rays was merely a charged helium atom, missing all its electrons? He set out to prove that this hypothesis was a sound one, and put a sample of a radioactive element in a glass tube that had walls thin enough for the alpha particles to penetrate. Therefore, the alpha particles could shoot through the glass of this tube. But the tube was situated inside a larger tube that contained no air and only a vacuum—at least as perfect a vacuum as could be achieved by the technology of the time. The glass walls of this tube were thicker and could not be penetrated by the alpha particles. Thus the particles could not escape from the larger tube. After the alpha particles had been coming from the inner tube into the outer one for some time, Rutherford noticed that a light gas was accumulating in the larger tube. Chemical tests proved that this gas was helium. But all that could have entered the outer tube were the alpha particles from the inner one. This experiment proved that alpha particles must be nuclei of helium atoms.

Rutherford also investigated the nature of radioactivity. He and another chemist showed that the radioactivity of radium, for example, consisted of the radium atom breaking down into an alpha particle and an atom of the heavy radioactive gaseous element radon. The alpha particle pulled electrons from its surroundings and finally became a neutral helium atom. In other words, radioactivity involved one chemical element changing into others—a process that nineteenth-century chemists and physicists believed could not happen. This finding constituted the final proof that the atom consists of smaller parts. It meant that atoms could sometimes break into lighter ones.

The beta rays, or electrons, emitted in radioactivity also had to come from within the atom. So the cathode ray phenomenon, the photoelectric effect, and radioactivity all indicated that the atom had an inner structure, and that the electron, the smallest unit of negative charge, was one of its parts.

THE ATOMIC NUCLEUS

What was the smallest unit of positive charge? And how did various atomic nuclei differ? Since electrons are very light particles, the atomic nucleus not only had to account for all the positive charge in the atoms, but also for over 99.99 percent of its weight. It was known that hydrogen and helium atoms were the lightest atoms. Rutherford had shown that the alpha particle was the nucleus of the helium atom. The particle also had a charge twice that of the electron in size. So two electrons would have to be added to an alpha particle to make a neutral helium atom, and it was concluded that the helium atom must have two electrons. The only lighter atom was that of hydrogen. No one had ever found a fraction of an electron. Only whole electrons and whole numbers of them can exist. Also it seemed logical that a lighter atom should contain fewer electrons than a heavier one. So, if the helium atom contained two electrons, it was logical to conclude that the hydrogen atom had one electron. But that meant that the nucleus of the hydrogen atom had a positive charge equal in size to the negative charge on the electron; that is the only way a neutral hydrogen atom can be formed. The hydrogen atom had one fourth the mass of the helium atom. So the hydrogen nucleus had to be about one fourth as heavy as the alpha particle, or helium nucleus. But then why was its charge not one fourth that of the alpha particle? After all if the atomic nucleus was the seat of the positive charge in the atom, why did not the weight of the atomic nucleus double when its charge doubled? In the case of electrons, for example, two electrons had twice the charge of one and also twice the mass of one electron. Why was the helium nucleus four times as heavy as the hydrogen nucleus, which had half its charge, instead of twice as heavy? How did the two nuclei differ? If the hydrogen nucleus was taken to be the smallest unit of positive charge, the helium nucleus had to contain two such units, according to its charge. But then why was it not twice as heavy as the hydrogen nucleus? Questions like these were not answered for two decades after Rutherford's experiments in 1911.

Hydrogen nuclei were isolated in experiments making use of modification of the ordinary cathode ray tube. Instead of a plane cathode, the tube used had a cathode that contained many small holes or canals in it. Then, as cathode rays went from cathode to the anode of the tube in the usual way, rays of a somewhat different nature were seen to emerge from the canals in the cathode on the side opposite the anode. The particles of the rays proved to have a positive charge while their mass and the size of their charge depended on the nature of the gas in the tube. If the gas was hydrogen, the particles had a mass about equal to that of the hydrogen atom and a positive charge equal in size to that on the electron. In this case, the canal rays had to be a beam of moving hydrogen nuclei. The particles of the hydrogen canal rays became known as *protons*, and are the smallest units of positive charge in ordinary matter.

This meant that the hydrogen atom was made of a proton at its center orbited by a single electron. But in this view the helium nucleus should hold two protons, and thus should weigh twice as much as the proton. Why did it weigh four times as much? The problem existed with other atoms. Other experiments had determined the charge on the nuclei of heavier atoms. The carbon nucleus had a charge six times as much as that on the proton, and therefore had to have six orbital electrons around it to make a neutral carbon atom. But the carbon atom weighed about 12 times as much as the hydrogen atom. Therefore, its nucleus had to be about 12 times as heavy as the proton. But its charge indicated it should hold six protons and not 12. The oxygen atom had a nucleus with a charge about eight times that of the proton, yet it had a weight about 16 times as much as the hydrogen atom. This all meant that the oxygen atom had eight orbital electrons. Again the charge and mass of the nucleus did not seem to agree with each other. Now it seemed logical to believe that the atomic nucleus must contain a number of protons. How else could the weights of most atoms be almost whole number multiples of that of the hydrogen atom? But why did nuclear charge measurements and atomic weight measurements disagree?

At first it was thought that all nuclei except those of hydrogen contain both electrons and protons. Then the helium atom's nucleus would consist of four protons and two electrons. That would give it a charge twice that of the protons and a weight about four times as much as it was known to have. The carbon nucleus would hold 12 protons and six electrons. That would give it the mass and charge it was known to have, while the same would hold for other nuclei in this electron-proton model of the atomic nucleus, and the theory did seem to account for the mass and charge of atomic nuclei.

But it at first seemed to do a lot more. It had been shown that the radiation from radioactive elements came from the atomic nucleus. In radium, for example, the heavy radium nucleus was seen to spontaneously emit an alpha particle, leaving behind a nucleus of the radium atom. But electrons, the beta rays of the radioactive process, were also emitted by many radioactive elements, and the electron-proton model of the atomic nucleus seemed to account for this easily. Nuclei of these elements were simply giving off some of the electrons they contained.

The model also solved a problem that would be present if the nucleus was seen to consist of protons alone, and that would be present even if the contradictions between the mass and charge of nuclei did not exist. It can be summed up in the question: What would hold a group of likely charged protons together in the nucleus? The electrical repulsion between them would be so strong that they would fly apart almost instantly. Only hydrogen atoms, with nuclei of single protons, would exist. What accounts for the existence of heavier atoms? The electron-proton model seemed to do so. The electrons and protons of the nucleus are oppositely charged and attract each other, and the attraction would hold nuclei heavier than those hydrogen together. In very heavy nuclei like those in uranium and radium the number of likely charged protons was large. As a consequence these nuclei were envisioned as not being very stable, since the repulsion between the large numbers of protons packed close together in a large nucleus would tend to disrupt such nuclei. Therefore, they sent out groups of electrons and protons, the alpha particles, and single electrons.

The electron-proton model of the nucleus seemed perfect at first. Yet it was doomed to failure. By the late 1920s physicists had worked out the laws of motion that governed subatomic particles like electrons and protons. These proved to be quite different from the laws of motion that applied to large bodies of daily experience, and had one grave consequence for the electron-proton model of the nucleus. These new laws implied that as an electron or proton is confined to a smaller and smaller space, its energy of motion increased greatly—an effect that became more dominant the lighter the subatomic particle is. Calculations showed that the effect would not be great for the heavier protons of the nucleus, but would be very great for the light electrons confined to the small space of the atomic nucleus. The electrons would be moving so fast under these conditions that they would fly out of the nucleus even against the strong electrical attractive forces of the protons. So the nucleus should yet fly apart in the electron-proton model, just as if it was made of protons only. In 1930 nuclear physicists were perplexed by this problem. Was there any solution to the thorny problem of atomic structure? All evidence indicated that protons had to exist in the atomic nucleus. But what held them together? Also what could account for both the charge and mass of nuclei if they were made, at least in part, of protons?

DISCOVERY OF THE NEUTRON

In 1932 a discovery was made that finally led to a solution of this puzzle. In that year an English physicist James Chadwick was engaged in experiments that involved bombarding the light metallic element beryllium with alpha particles from a radioactive source. In these experiments, a sheet of paraffin was placed in front of the beryllium sample bombarded. Under alpha particle bombardment, Chadwick noticed that the beryllium sheet gave off a strong radiation. When the radiation fell on the paraffin sheet, it could knock protons out of it. Now paraffin contains molecules holding only carbon and hydrogen atoms. It was shown that the protons emitted came from the hydrogen atoms in these molecules. The strange radiation from beryllium exposed to alpha particles, whatever it was, had the ability to knock the protons out of the hydrogen atoms in the paraffin molecules. After many considerations, Chadwick postulated that the radiation from beryllium was made of electrically neutral particles that had about the same mass as protons, which became known as *neutrons*. Their existence was soon verified by other experiments.

But how did the neutrons arise in Chadwick's experiments? Recall that the experiment made use of alpha particles. The only possibility was that the moving alpha particles collided with the nuceli of beryllium atoms as they moved through the mostly empty atoms of the light metal. And when they did so, they must have knocked parts out of these nuclei which could only have been the neutral particles (or neutrons) that knocked protons out of the paraffin sheet. This finding indicated that the atomic nucleus must also contain neutrons as well as protons. So the proton-neutron model of the atomic nucleus was developed.

This model solved the charge and mass problems that had plagued the electron-proton model. In this model the hydrogen nucleus was made of a single proton, while the alpha particle, or helium nucleus, is made up of two protons and two neutrons. That accounts for its positive charge of twice the size of that on the electron, since the neutrons are electrically neutral. It also accounted for the alpha particles' weight. The particle is composed in this model of four particles of about the same mass as the proton. Therefore, the particle would weigh about four times as much as the proton, or hydrogen nucleus. The carbon nucleus contained six protons and six neutrons. The six protons accounted for its charge of six times that of the electron in size. It contained 12 particles altogether, giving it a weight of about 12 times that of the proton. The same considerations applied to other atomic nuclei.

But if the model solved these problems well, it, at first, seemed to present others. One of these was the old one of what held the likely charged protons of the nucleus together. The neutrons were neutral and could not do so. Another was the origin of the electrons emitted by radioactive elements. It had been shown by experiment and theoretical considerations too involved to present here that these electrons had to come from the atomic nucleus. But from where in the nucleus of the proton-neutron model could they come? There were no electrons in the nucleus in this model. But the situation had again become the same as with the Rutherford model 20 years earlier: experiment clearly showed that protons and neutrons exist in the atomic nucleus.

It was speculated in 1935 and later confirmed that a strong nonelectrical force acted between protons and neutrons that bound them together in the atomic nucleus. It became known as the *nuclear force* or *strong force*. It is a very short-range force. That is, it is many times stronger than the electric force between two protons when the two particles are less than a diameter apart, but when they move more than a diameter apart, the force drops to almost nothing. This behavior provided a clue as to why alpha particles were emitted from nuclei of heavy radioactive elements. These nuclei contain many protons. They repel each other by the electric force. But the electric force is long-range. This means that it remains strong when two protons are many diameters apart. Because of this, every proton in a large atomic nucleus feels the repulsion of all the others. But, because the nuclear force is short range, every neutron or proton in the nucleus just feels the attraction of its nearest neighbors, and not that of all the particles of the nucleus. Though this attractive force is much stronger than the electric force, the fact that it is short range makes large nuclei unstable, so that they emit fragments made of protons and neutrons, or alpha particles, at high speed. So the proton-neutron model accounted for the emission of alpha particles from nuclei. But what accounts for the electrons emitted from nuclei of radioactive elements?

This question will be addressed shortly, after the details of the electron structure of the atom is presented. If this aspect of atomic structure is given at this stage it will not only unify our account of the subject, but will also prepare the way for topics in later chapters dealing with the structure of proteins and nucleic acids. Also, the emission of electrons from radioactive elements gets to the heart of our topic in this section: radioactive isotopes. That subject is best presented just before we discuss their use in biochemistry.

The electrons that orbit the atomic nucleus do not do so helter-skelter; there is much order in their motion. But, to understand it in detail, it would be necessary to delve deep into mathematical physics, which is out of the question here. So only the overall details can be presented.

THE BOHR ATOM

Why did the electrons not lose energy and fall into the nucleus, as a classical electromagnetic theory said they should? The great Danish physicist Niels Bohr pondered this contradiction in 1913, two years after the Rutherford model of the atom was proposed. He first considered the simplest case of the hydrogen atom. This atom is made up of one proton orbited by a single electron. Bohr postulated that the electron in this atom could orbit the proton only in circular orbits at certain distances from the proton. It could not orbit the proton at any distance between these fixed distances. In each such orbit the electron had a fixed energy. So the electron in the hydrogen atom could have only certain energies, and no others between them. In its normal state, the electron orbits the proton in the circular orbit nearest to the nucleus in this model of the atom. In this orbit it has its lowest energy. In all orbits at a greater distance than this from the nucleus, the electron has a greater energy. However, in Bohr's theory, the state of lowest energy, the orbit nearest to the nucleus, is the stable state for the electron and it cannot fall toward the nucleus in this state.

Bohr had more than the observed stability of atoms to back up his theory. In fact, an important piece of experimental evidence showed that his model of the atom had to be valid. It involved the light emitted by gaseous elements when an electric current is passed through them. The gas in the tube in this case must be in atomic rather than molecular form; that is, it must consist of single atoms not joined into molecules. Also there must be at a little more gas in the tube than when cathode rays are produced. But the gas must be at a lower pressure than ordinary gases like air. Then, when an electric current is passed through the tube, the tube fills with a glow of a certain color. The color depends on the gaseous element in the tube. When the light emitted by the gas is analyzed with a *spectroscope*, an instrument that splits it into its component colors or wavelengths, the light was found to consist of only certain colors of fixed wavelengths. The light emitted by such gases in atomic form consists of only certain wavelengths, and not the continum of wavelengths of normal white light. If, however, the electron was seen to fall into the nucleus in the way classical electromagnetic theory implied, it should emit all wavelengths in doing so, not only fixed wavelengths.

Bohr also interpreted these results in terms of another theory of radiation that gained support in 1905. By that year the photoelectric effect had been thoroughly studied. Some peculiar phenomena were found that could not be explained by the wave theory of electromagnetic radiation. First, the energy of the emitted electrons did not change as the intensity of radiation incident on the metal was increased. If the radiation was made of varying electric and magnetic forces, an increase of intensity of the radiation meant an increase in the strength of these forces. The stronger forces should then act more strongly on the electrons at the surface of the metal. Thus the electrons should be boosted from the metal with more speed or energy as the radiation intensity is increased. But they were not. The average energy of the emitted electrons remained the same, but more of them were emitted as the intensity of the incident radiation was increased. Why? Secondly, above a certain wavelength, for a given metal, no electrons would be emitted by the radiation no matter how high its intensity. By the wave theory, some electrons should have been emitted at all wavelengths of the incident radiation, even at the lowest intensities. This is because the electric forces comprising the radiation would still act on the electrons at the metal's surface and would finally give them enough speed to escape from the metal. The wave theory had no explanation for these two peculiarities of the photoelectric effect.

PHOTONS

In 1905, the great physicist Albert Einstein showed that these aspects of the photoelectric effect could be explained if electromagnetic radiation is seen as consisting of little particle-like packets of energy called *photons*. He assumed that a photon of the incident radiation collided with an electron at the surface of the metal. In the process the photon would transfer its energy to the electron. Since the photon had only so much energy, the electron could leave the metal with only a certain maximum speed or energy. If the intensity of the radiation was increased, the number of photons incident on the metal increased. This, in turn, meant that more such electron-photon collisions occurred. Thus more electrons were emitted, although their average energy did not change. Einstein also assumed that the energy of the photon decreased in direct proportion to an increase in the wavelength of the radiation. That is, if the wavelength was doubled, the energy of the photon is halved. The electrons at the surface of the metal are bound there by electric forces. When a photon collides with such an electron, it must free the electron against these forces that bind it in the metal if the electron is to leave the metal. If the wavelength of the photon—that is, the wavelength of the radiation the photon comprised—was above the maximum wavelength for which the photoelectric effect would occur for a given metal, the photon did not have enough energy to do this, and no electrons would be emitted, no matter what the intensity of the radiation.

Einstein had shown that a particle model of electromagnetic radiation could explain the photoelectric effect. Did that mean that the wave theory of radiation was wrong? No. Both the particle and wave models were needed. In fact Einstein could not have talked about wavelengths of the radiation if the radiation did not have a wave aspect. The photon model explained events that the wave model could not. And the wave model explained other events the photon model could not. Physics had known no other situation like this—two different theories explaining the same phenomenon equally well. But it had to be accepted. Electromagnetic radiation came to be seen as made of moving photons that are guided along their paths by a wave field.

Bohr used the photon model of light in his theory of the atom. It helped explain why the light and other electromagnetic radiation emitted by atoms of a gas in a cathode ray tube came only in certain wavelengths and no others. Say the gas in the tube is atomic hydrogen. Normally the orbiting electron in each of the hydrogen atoms would be in its orbit of lowest energy nearest to the proton. But, when the current is turned on, electrons move through the tube at great speed from the cathode to the anode. These moving electrons often collide with hydrogen atoms as they go from cathode to anode. These collisions can knock the electrons in these atoms into another orbit of higher energy. These orbits are less stable than that of the lowest energy. So the electrons fall back to this level almost immediately. Or some might fall first to the next orbit of lower energy just below its present energy, and then to the next, and so on, until they reach the orbit of lowest energy nearest the proton. Each time the electron falls from an orbit of higher energy to one of lower energy, it emits a photon of radiation. The energy of the photon emitted is equal to the difference in energy between the two orbits. Since the energy of each photon emitted is fixed, the wavelength of the radiation it corresponds to must also be fixed. That is why the light emitted from the gas in the tube was of fixed wavelengths only, and no others. This experimental fact, as well as the stability of the atom, spoke for the validity of Bohr's atomic model.

QUANTUM MECHANICS

This atomic model, too, was modified as physicists discovered the laws of motion of subatomic particles in the 1920s. In 1924 the French physicist Louis de Broglie postulated that particles like electrons and protons also display a wave nature like light waves displayed a particle aspect. His theory was soon proved true. That development led to the formulation of a branch of mathematical physics called *quantum mechanics*; it dealt with the laws of motion of subatomic particles. It was calculated on basis of quantum mechanics that electrons could not exist in the nucleus of the atom. In quantum mechanics, the smaller the space to which a particle is confined, the larger becomes its speed or energy, and when the same kind of calculations were done for neutrons, their energy in the nucleus was shown to be ideal for their existence in the nucleus. They move much slower than electrons would under the same circumstances. Thus they stay in the nucleus. However, the wave nature of particles that lies at the basis of quantum mechanics had implications for the electron structures of atoms also. The ability of electrons to exist in only certain energy states in the atom is a result of quantum mechanics, though the atom based on quantum mechanics is a little different than the early Bohr model. In that model one could talk about definite electron orbits. In the new model one cannot. Here the electron has a wave nature as well as a particle one, and because of that, it can be shown that the electron's position and speed cannot be measured at the same time, which meant that the electron cannot really be visualized as moving in a specific orbit. All that one can speak of are certain regions about the nucleus in which the electron of the hydrogen atom is likely to spend most of its time. These regions correspond to the Bohr orbits in the older model.

ELECTRON SHELLS

The Bohr model of the atom also had difficulties when it came to the electron structures of atoms heavier than hydrogen. Here again, the atomic model based on quantum mechanics was successful. Each of the regions about the nucleus in which the electron of the hydrogen atom can spend most of its time are called *electron shells*. In its lowest energy state, the electron of the hydrogen atom is in the shell of lowest energy nearest to the nucleus which is known as the *K shell*. This shell can hold up to two electrons at most. The next atom, that of helium, has two electrons. Both of them go into the K shell. So that shell is full in the helium atom. The next several atoms in order of increasing weight, along with the number of electrons and protons they contain, are shown in TABLE 8-1. Since all these atoms have more than two electrons, their K shells are full. The next

Table 8-1
Some chemical elements
and the number of
electron-proton pairs

Atom	*Symbol*	*No. of electrons*	*No. of protons*
Lithium	Li	3	3
Beryllium	Be	4	4
Boron	B	5	5
Carbon	C	6	6
Nitrogen	N	7	7
Oxygen	O	8	8
Fluorine	F	9	9
Neon	Ne	10	10
Sodium	Na	11	11
Magnesium	Mg	12	12
Aluminum	Al	13	13
Silicon	Si	14	14
Phosphorus	P	15	15
Sulfur	S	16	16
Chlorine	Cl	17	17
Argon	Ar	18	18

shell, the *L shell*, just beyond the K shell out from the nucleus, can hold up to eight electrons. So the lithium atom has two electrons in the K shell and one in the L shell. The carbon atom, with six electrons, has two in the K shell and four in the L shell. The nitrogen atom, with seven electrons, has two in the K shell and five in the L shell. The oxygen atom, having eight electrons, has two in the K shell and six in the L shell. The neon atom has ten electrons. It therefore has two electrons in the K shell and eight in the L shell. So in the neon atom both the K and L shells are full.

The sodium atom has 11 electrons. Two of these go into the K shell, eight into the L shell, and one into the next shell out from the nucleus, the *M shell*, which can hold up to 18 electrons in heavier atoms. A silicon atom has 14 electrons. Two of these go into the K shell. Eight into the L shell. The remaining four go into the M shell. The argon atom has 20 electrons. Two of them go into the K shell, eight into the L shell, and eight into the M shell. In lighter atoms like that of argon, the M shell is full at eight electrons, though this shell can hold up to 18 in heavier atoms. Figure 8-5 shows the shell structures of four common atoms.

The heaviest known atoms have up to seven electron shells. The shells, and the maximum number of electrons each can hold, are shown in TABLE 8-2. No electron shell holds more than 32 electrons in known atoms. Known atoms do not have enough electrons to supply a shell with any more. This excursion of atoms is important in regard to some aspects of protein and nucleic acid later.

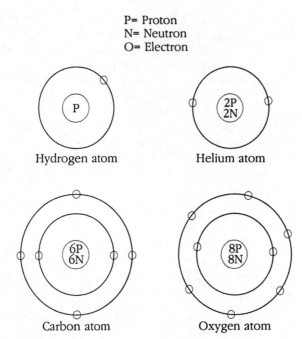

P= Proton
N= Neutron
O= Electron

8-5 Atomic shell structures for common elements.

Table 8-2   The shells and the maximum number of electrons

Shell	No. of electrons
K	2
L	8
M	18
N	32
O	50
P	72
Q	98

ISOTOPES

In the first 20 atoms just discussed, the number of neutrons in their nuclei was not mentioned, for a good reason: the number of neutrons they hold varies. The number of protons in the nucleus always equals the number of electrons in neutral atoms. But the number of neutrons can differ in atoms of the same element. Take hydrogen. Most hydrogen atoms have nuclei of one proton. But some—a small minority—have a nucleus of one proton and one neutron. This kind of hydrogen is called heavy hydrogen or *deuterium*. Deuterium atoms are about twice as heavy as ordinary hydrogen atoms. Varieties of the same element having atoms with different weights, because of different numbers of neutrons in their nuclei, are known as *isotopes*. Hydrogen has a third isotope that is even rarer than deuterium. It is called *tritium*. Its nucleus has one proton and two neutrons, and is therefore about three times as heavy as ordinary hydrogen. All three hydrogen isotopes have one proton in their nuclei, and hence one orbital electron. Therefore, they all have the same chemical properties.

Helium has two isotopes. One of these is the normal form of helium, having two protons and two neutrons in its nucleus. This isotope is known as *helium-4* or He-4. A rare isotope of helium has two protons and one neutron in its nucleus. It is called *helium-3* or He-3. The numbers refer to the mass number of the isotope. The mass number of an isotope is the total number of particles in its nucleus, or the sum of the number of protons and neutrons. For normal helium the mass number is 4, while for the lighter isotope of the element it is 3. Carbon has three isotopes. One of these, the most abundant, has six protons and six neutrons in its nucleus, and therefore has a mass number of 12. It is represented by the symbol C-12. Another isotope of carbon has a mass number of 13, and has six protons and seven neutrons in its nucleus. It is represented by the symbol C-13. This carbon isotope is not very abundant in nature. An even rarer isotope of carbon, C-14, has a mass number of 14. It has six protons and eight neutrons in its nucleus. Nearly all elements have two or more isotopes, one of which is usually the most abundant while many of the others often come in traces.

The discovery of isotopes shed light on why electrons were emitted by some radioactive elements.

There were no electrons in the nucleus of the proton-neutron model, but in some isotopes, the number of neutrons in the nucleus exceeds the number of protons a little too much for the nucleus to be comfortable. In the mid-1930s it was postulated by some noted physicists that one of the neutrons in such a nucleus transforms itself into a proton, an electron, and an elusive photon-like neutral particle known in modern physics as the *antineutrino*. The electron formed flies out of the nucleus in the form of beta radiation, as does the antineutrino, while the proton formed stays in the nucleus. The nucleus thus has become one of another element. The element formed contains one more proton in its nucleus than the original element undergoing *beta decay*.

Consider the three hydrogen isotopes once again. Ordinary hydrogen and deuterium are stable isotopes. The deuterium nucleus has one proton and one neutron, and therefore the proton-neutron balance is perfect in this nucleus, and the neutron has no tendency to change to a proton. But with tritium, or H-3, the situation is different. The tritium nucleus has two neutrons and one proton, so that there are twice as many neutrons as protons. Thus one of the neutrons changes to a proton, emitting an electron from the nucleus, which in the process has become a nucleus containing two protons and one neutron, and is a nucleus of helium-3. Tritium is an important radioactive isotope in radioactive tracing. In the case of carbon,

C-12 and C-13 are stable isotopes, while carbon-14 is radioactive, since the carbon-14 nucleus has six protons and eight neutrons, which is a little too many neutrons. One of them changes to a proton, emitting an electron. This leaves the resulting nucleus with seven protons and seven neutrons, making it a nucleus of the most stable isotope of nitrogen, N-14. Nitrogen has two stable isotopes—N-14 and N-15.

In some cases the number of protons in a nucleus exceeds the number of neutrons. If the excess is too much, one of the protons transforms itself into a neutron. In the process it emits a particle of the same weight as the electron, but with a positive charge, called a *positron*, and a photon-like particle called a *neutrino*, both of which leave the nucleus.

With the advent of the nuclear reactor, beams of neutrons became available. Such neutron beams could be fired at targets made of the most stable isotope of an element, and the nuclei in the target then capture neutrons of the beam and become nuclei of another isotope of the element. Many of these man-made isotopes are radioactive. And after the nuclear reactor became available many such radioactive isotopes of lighter elements could be produced.

RADIOACTIVE TRACING

This introduction to atomic structure and isotopes leads to discussion of the process of *radioactive tracing*. The process is based on the fact that a radioactive isotope of an element has the same chemical characteristics as the normal (most abundant) form of the element. Therefore, if the element is one needed by a living organism, its radioactive form, when supplied to the organism in its food, will be taken into the same biochemical processes as the normal nonradioactive form of the element would be. The most commonly used radioisotopes in biochemistry are the *beta emitters*, or those that emit beta rays or electrons. When a radioactive element is supplied to an organism, it can be traced through the organism's body because of its radioactivity by the use of a *gieger counter* or *scintillation counter*, the most common devices used to detect radioactivity. A radioactive isotope used to examine the biochemical processes of an organism is an example of a *radioactive tracer*. Radioactive tracers found great use in breaking the genetic code.

The Hershey-Chase experiment made use of radioactive isotopes of phosphorus and sulfur. The radioactive phosphorus was taken into the DNA cores of the virus and thus got into the infected cells. The radioactive sulfur got into the protein coats of the viruses. It was therefore left in the liquid outside the infected cells. Both of these radioisotopes are beta or electron emitters. But the energy of the electrons emitted in each case is different. A detector like a scintillation counter not only detects these electrons; it also measures their energy. Thus Hershey and Chase could tell whether the emitted beta rays were those of radioactive sulfur or those of radioactive phosphorus by the energy of the electrons emitted.

Isotopes of nitrogen were used in the Meselson-Stahl experiment, although they were not radioactive. Yet the procedure of the experiment still demonstrated how isotopes can be used in biochemistry. In that experiment, the difference between the two nitrogen isotopes—N-14 and N-15—was not radioactivity but a difference in weight. The centrifuge was employed to detect this difference.

Most isotopes used in biochemical studies are radioactive. Radioactive tracing is one of the chief modern methods of biochemistry and genetics, and is used in conjuction with chemical methods, the ultracentrifuge, chromatography, and X-ray studies to answer many basic questions about the genetic code and the chemistry of the cell.

THE ELECTRON MICROSCOPE

At this point one other research tool of the present century will be discussed briefly: the *electron microscope*. The electron microscope, first developed in the 1930s, made cell organelles visible with much more efficiency than any other method of microscopy.

The electron microscope was the best solution to the problem caused by the small size of cell organelles. These objects are for the most part too small to scatter the relatively large waves of visible light effectively. The electron microscope has a much greater *resolving power* than the light microscope. The resolving power of a microscope is simply the minimum distance two objects must be separated by, in order for the microscope to make them visible as separate objects. The smaller the resolving power of a microscope, the more detail the microscope can reveal when viewing small objects. For the electron microscope the resolving power is much smaller than for the light microscope. Consequently, when the instrument was developed in the 1930s, cytologists could see many cell organelles that had been invisible or unclear under the light microscope. When technical improvements were made on the electron microscope after its invention, all organelles of the cell became visible.

The electron microscope uses moving electrons instead of electromagnetic radiation as a means to probe small objects. Electrons also have a wave nature besides a particle nature, and that is the basis of the operation of the electron microscope. The wavelengths of the electrons used at moderate energies is much smaller than those of visible light, and the electron waves of a beam of electrons under these conditions are very small compared to the size of cell organelles. Thus the electron waves are easily scattered by them, revealing much more detail about the structure of the cell than optical microscopes.

Again the design of this microscope is a branch of technology that cannot be pursued here. Only basic principles can be given. The important point is that the electron microscope utilizes a beam of electrons that can be given different energies and thus different wavelengths. At large energies the small wavelengths of the electrons of the beam can reveal the details of the structure of various organelles. At even smaller wavelengths of the electron beam, the microscope can make many biomolecules visible.

Electron waves are invisible to the human eye. However, the electrons of the beam darken photographic plates. After the electron beam is scattered or reflected from the small objects being viewed, the micro-scope directs them to a photographic plate. When the plate is chemically developed, it gives an image of the object being viewed in various shades of darkness. Such an image is known as an *electron micrograph*.

Chapter 9

Breaking the genetic code

Only the overall procedures of the work of Avery and his team were presented in Part I of this book. Nothing was said about how he obtained solutions of pure nucleic acid that caused transformation of the rough avirulent pneumococci bacteria to the smooth virulent kind. Actually, it all started by following clues given by Griffith's work in 1928. Recall that the microscope revealed that the blood of mice given an injection of a solution of heat-killed, virulent bacteria and avirulent ones was teeming with live virulent pneumococci. Avery and the scientists working under him followed up on this observation in the 1930s.

A CLOSE LOOK AT AVERY'S WORK

In the early 1930s, they showed that such transformation of the pneumococcus bacterial strain could occur outside the body of the living organism. A solution of heat-killed, smooth pneumococci could easily be obtained. If some live rough pneumococci were put into this solution, it was seen that the live rough bacteria developed into the smooth type having the carbohydrate coat in many cases. They were thus transformed into the smooth type of bacteria by merely being placed in a solution containing heat-killed disease-causing bacteria. This was only the start of the productive work to come out of Avery's laboratory at the Rockefeller Institute in New York in the 1930s.

After Avery's team had shown that transformation could take place outside the living organism, the question of what caused the transformation became all the more important. The most reasonable suggestion was that the heat-killed, smooth pneumococci passed some gene into the hot water used to kill them. This gene governed the formation of the carbohydrate coating of the disease-causing pneumococci. Such a gene had to be contained in some substance given up by the heat-killed virulent cells, which was picked up by the avirulent rough cells when they were placed in a solution of the heat-killed smooth cells. But how could the substance be isolated from the solution?

Much headway in doing this was made by James Alloway, one of the scientists working under Avery in the early 1930s. In order to really make progress in isolating the transforming substance, a more concentrated solution of the substance would have to be obtained. When you merely take a colony of smooth pneumococcus cells and put them into hot water, most of them die. Few of them are disrupted so that they can pour their contents into the solution. So Alloway went farther than this. He first took a large colony of disease-causing, smooth pneumococci and killed them by putting them into hot water, and then cooled the solution until it froze, after which he heated it again. It was found that such freezing followed by heating of the solution of dead cells had the effect of breaking the cells apart. Thus most of the cells burst open on such treatment and spilled their organelles and component substances into the solution. In this way, Avery could obtain a more concentrated solution of the gene that governed the formation of the carbohydrate coating of the disease-causing pneumococci, and hence, a solution more concentrated in the substance that held this gene. But the solution held many other components and disrupted cell walls, as well as many other substances dissolved in the solution in addition to the genetic material that Avery and his team were looking for.

However, Alloway made much progress in getting a solution more pure in the genetic material, DNA, and in this notable feat he made use of the ultracentrifuge. He took a tube of the complex solution holding many organelles and dissolved substances, and spun it at great speed in a powerful centrifuge. As the tube was spun, the broken cell walls and organelles settled near the bottom of the tube, while the clearer portion of the solution holding the genetic material and other dissolved substances stayed near the top of the tube. He separated this part of the solution from that at the bottom of the tube containing the organelles and cell walls. The mixture he had gotten in this way was richer in the genetic material than others had been, and was therefore very effective in transforming rough pneumococci into the smooth ones. Alloway's work was a breakthrough for the Avery team. Yet the solution contained many other dissolved substances in addition to the genetic material, and at that time, it was not even certain in many circles that DNA was the genetic material, although there was increasing speculation that it might be. What was now needed was a means to obtain a pure solution of the genetic material whatever it might be.

In the mid-1930s, most experts believed that proteins were the genetic material. Was the active transforming substance in the solution that Alloway obtained a protein or nucleic acid? Avery decided it was time to find out one way or another.

He took charge over the work going on in his laboratory and decided that he would do more intricate experiments with the solution Alloway had gotten from heat-killed, smooth pneumococci through use of the centrifuge. He did not perform these experiments alone. McCarty and MacLoed aided in them, as explained in chapter 3. The first thing Avery did was to put a colony of the live, smooth pneumococci into a salt solution. The solution had the effect of keeping the cells suspended and separated from each other in the solution. He then heated the salt solution to kill the cells and inactivate any enzymes that might be present that had the ability to destroy the transforming substance, the genetic material. Some such enzymes were known, so they had to be guarded against in the experiment. He then treated the dead cells with additional salt solution. The next step was to dissolve the cell membranes of the dead bacteria, since it would be better if they could be dissolved rather than merely disrupted as Alloway had done; then there would be less solid material to centrifuge out of the mixture. That would be less time-consuming and would make chemical analysis easier.

But how could the cell walls be dissolved? To do this Avery and his team used a solution of the substance sodium desoxycholate. This substance had the effect of dissolving the cell membranes and breaking apart the carbohydrate coatings of the dead cells. When a significant amount of sodium desoxycholate solution was added to the salt solution holding the dead cells, and the resulting mixture was shaken, the cell walls of the dead pneumococci were dissolved, and the cells spilled their contents into the solution. The resulting mixture held various cell organelles, carbohydrates, proteins, nucleic acids, and the disrupted cell walls. The mixture was then centrifuged to separate the fragments of the disrupted cell walls and all organelles from the rest of the mixture. After these solid components were separated off, Avery had to decide how to deal with the rest of the mixture to see which of its components was the genetic material.

At this point he made use of a liquid substance, *ethanol*, a simple alcohol, in his chemical procedures. DNA was known to be insoluble in ethanol to a great degree; that is, DNA would not mix with or dissolve in the liquid, while a lot of other cellular substances like proteins, carbohydrates, and so on, did dissolve in it to various degrees. So, at this point, Avery added a large volume of ethanol to the complex mixture he obtained from the dead smooth cells. That had the effect of causing an amount of a solid substance—mainly DNA and some proteins and carbohydrates tightly bound to it—to separate out of the solution and go to the bottom of it. But the solid material still was much richer in the DNA than the original solution had been, while the sodium desoxycholate was left behind in the liquid part of the solution. Avery then poured the liquid over filter paper. The liquid portion of the mixture leaked through the paper while the solid portion was left behind. In this way Avery had separated the solid portion of the solution rich in DNA from the liquid portion, while at the same time getting rid of the sodium desoxycholate in the material. But the solid material was not yet pure DNA. Avery had to go further.

He next made use of a simple liquid substance known as *chloroform*, but first dissolved the solid material in salt solution. To the resulting solution he added a large volume of chloroform, which had the effect of separating the bound proteins from the DNA in the dissolved solid material. He had to shake the mixture vigorously to accomplish this. Now he separated the chloroform from the mixture, and again added ethanol to separate the DNA out once again. The resulting solid material was yet richer in DNA than the previous solid material still had been. The proteins had been separated from it by the chloroform treatment. Yet the resulting solid material held a detectable amount of carbohydrate substances.

To separate these from the DNA, Avery dissolved the solid material in a salt once again, and added an enzyme to the solution that breaks carbohydrates into simple sugars that easily separated from the DNA and went into the solution. Then ethanol was added to the solution to separate the DNA out as a solid. This solid was separated from the liquid by using filter paper. The resulting DNA, though very pure, still contained a lot of the enzyme used to break down the carbohydrates. But this enzyme, like all enzymes, was a protein, so that all that Avery had to do now was to redissolve the solid in salt solution and add chloroform to it to break down the enzyme. After the chloroform was separated from the resulting mixture, Avery had a solution of mostly DNA, which could easily be separated from the mixture.

It has been outlined here how Avery had gotten very small amounts of DNA of almost 100 percent purity. A solution of the DNA he obtained could transform rough pneumococci into the smooth type, showing that DNA was the genetic material in these experiments. A solution of pure proteins could not do the same.

It is important to note here that two chief methods of modern genetics played the leading role in this historic experiment: biochemical analysis and the centrifuge. Without them, such progress in isolating the genetic material could not have been made.

THE HERSHEY-CHASE EXPERIMENT

In the second historic development in modern genetics, the Hershey-Chase experiment, the methods of radioactive tracing and the centrifuge played the leading role. Recall that in this experiment, viruses having radioactive sulfur atoms in their protein shells and radioactive phosphorus atoms in their nucleic acid cores were allowed to infect bacterial cells in a growing medium. But in Part I, the impression given was that the nucleic acid cores just entered the cells while the protein coats just fell off the cells. Actually the matter was not that simple. The protein coats of the viruses, though they did not enter the cells, tended to stay bound to their membranes. If this situation could not be remedied, the cells and the viral protein coats would not have been able to be separated from each other. The experiment could not have been done if such was the case. Hershey and Chase got around this setback in an unexpected way. They found that the use of a simple food blender was the answer, and when a medium containing the infected cells was shaken moderately in the blender, microscopic studies showed that most of the viral protein coats were knocked off the infected cells without crushing them. The cells could not be crushed or disrupted in this experiment, since the investigators wanted them intact to see if they held radioactive phosphorus.

The use of the food blender illustrates an important point about how creativity operates in the progress of any science, including genetics. Active science is not restricted to any set methods within the realm of observation and measurement, although each science might have its chief ways of approaching problems. Individual scientists keep their eyes open at all times for new and unexpected techniques that might be of use in specific experiments. The use of an ordinary food blender in the Hershey-Chase experiment is a good example.

When Hershey and Chase had separated the infected cells and the protein coats of the viruses in the liquid growing medium of the cells, a way was needed to isolate the cells from the protein shells of the virus. Here again the centrifuge was put to use. When the liquid medium was spun at high speed in a centrifuge, most of the heavier cells were thrown to the bottom of the tube, while the protein coats of the virus stayed at the top of the tube. In this way, the investigators were able to separate the infected cells from the protein shells of the virus. They obtained two liquid portions from the original medium. One of these contained the infected cells, and the other the viral protein coats.

As explained in Part I, the investigators found that the infected cells contained radioactive phosphorus, while the liquid holding the protein coats of the viruses held radioactive sulfur. But both of these radioactive isotopes have an excess of neutrons in their nuclei, and are thus both electron emitters. But detectors could tell the investigators which isotope the electrons were emitted from through a difference in the energy of the electrons. In this way they found that most of the electrons emitted by the infected cells were those of radioactive phosphorus, showing that the nucleic acid cores of the viruses had entered the cells. On the other hand, most of the electrons from the liquid portion holding the protein coats of the viruses were those from radioactive sulfur, showing that the coats generally held no nucleic acid cores.

The separation of the two kinds of radioactivity was not really as clear cut in the actual experiment as some popular accounts portray. First, the food blender did not knock all the viral protein coats off the infected cells. The blender was not 100 percent effective in separating the protein shells from the infected cell's membrane in all cases. Secondly, the centrifuge is not perfectly effective in separating the components of a mixture of organelles and other cell parts. However, most of the radioactive phosphorus ended up in the infected cells, while most of the radioactive sulfur ended up in the liquid holding the protein coats, so that the experiment was considered conclusive in its chief results.

In the Hershey-Chase experiment, radioactive tracing and the ultra-centrifuge played the leading role; this was in 1952. By then it had been conclusively shown that DNA is the genetic material.

THE MESELSON-STAHL EXPERIMENT

The last historic experiment made possible by modern methods was the Meselson-Stahl experiment. It made use of two isotopes of nitrogen— N-14 and N-15. The first generation of bacterial cells descended from those having DNA with N-15 atoms in both chains were found to have N-14 atoms in one chain and N-15 atoms in the other. But how did the experimenters determine this? Well, after they had separated the DNA from a large number of the first generation cells, they subjected this DNA to a special kind of treatment in a powerful ultracentrifuge. It made use of a solution of the salt cesium chloride CsCl.

When a tube containing a concentrated solution of cesium chloride is spun rapidly in an ultracentrifuge, the salt concentrates more heavily toward the bottom of the tube. That is, the density of the solution in the tube increases steadily as one moves down the tube while the tube is being spun: a *density gradient* of the salt develops in the tube. Imagine now that the solution also holds many different cell organelles and large biomolecules; such a mixture can be put into a cesium chloride solution before it is spun in the centrifuge. Now since the solution has a different density at each point in the tube, each type of organelle or biomolecule will move to the place in the tube where the density of the salt solution equals its density. The type of centrifuge making use of this kind of salt solution is called a *density-gradient centrifuge*.

How did Meselson and Stahl make use of the process in their experiment? First they had to obtain a certain amount of DNA from the parent *E coli* which held DNA having the heavy N-15 atom in both chains. This kind of DNA should be the heaviest, since it holds the heavy N-15 atoms in both chains. After they had gotten this DNA, they placed it in a cesium chloride solution in the tube of a powerful ultracentrifuge and spun it at about 45,000 revolutions per minute. The DNA would then settle in a certain place of the tube where the density of the solution equaled its density, and formed a band in that area of the tube. But how could the experimenters detect the position of this band in the tube as it spun?

Here they made use of the fact that DNA absorbs ultraviolet radiation at a wavelength of 260 nm. If the solution in the tube was exposed to ultraviolet radiation of this wavelength, the radiation was absorbed only by the band of DNA at a certain point along the tube, while at all other points it was transmitted. The investigators used a special type of camera built to detect ultraviolet radiation to photograph the spinning tube. All parts of the tube that transmitted the 260 nm radiation would show up brightly on the film of the camera, except for the band of heavy DNA that absorbed the radiation. It would show up as a dark band at a certain distance down the tube. Ordinary DNA from the same kind of *E coli* cells that held only N-14 atoms in both chains should be less dense than that holding N-15 atoms in both chains. When such DNA was placed in the cesium chloride solution of the centrifuge tube, it settled as a band higher in the tube than that of the heavier DNA, as shown in FIG. 9-1. The experimenters noted the positions of the two dark bands of the heavy and light DNA. The band of the heavy type occurred further down the tube than that of the lighter DNA, since it was denser than the lighter type.

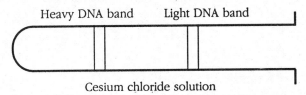

Heavy DNA band   Light DNA band

Cesium chloride solution

9-1 The separation of DNA in a cesium chloride solution after centrifuge.

But where in the tube would the DNA from the first generation of *E coli* settle? If that DNA held one chain of N-14 atoms and one chain of N-15 ones, its density should be somewhere between that of ordinary DNA and that of heavy DNA holding N-15 atoms in both chains. Thus it should settle at a point in the tube between the bands of the heavy and light DNAs. Meselson and Stahl found that such was the case. This offered the best proof that the first generation DNA in the experiment held one chain of N-15 atoms and one of N-14 atoms. That could only mean that DNA replicates semiconservatively, as was said earlier.

What kind of DNA did the experimenters find in the second generation of *E coli* cells grown in the N-14 medium by this method? When they isolated DNA from these cells, and subjected it to the density gradient ultracentrifuge, they found it consisted of two kinds of DNA. One of these settled in the same place in the tube as ordinary DNA does, which showed it must have been ordinary DNA holding two chains of N-14 atoms. That, again, is what would be expected if DNA replicates semiconservatively. The other component of the second-generation DNA settled in the same part of the spinning tube as the first-generation DNA did. This showed that it held one chain of N-14 atoms and one of N-15 atoms, while the proportions of the two DNAs (as indicated by the intensity of the darkness of their bands) was what was expected from the semiconservative method of DNA replication.

The methods of modern genetics and biochemistry played a major role in showing that DNA is the genetic material in the 1940s and 1950s. Such a revelation would have been impossible without them. By the late 1950s, the once mysterious genes of Mendel had been shown to be a sequence of base groups on the DNA molecule of the chromosomes. But these research methods have not stopped shedding light on the nature of the gene in recent years. Next they were applied to break the genetic code. That is, they were put to use to discover what particular triplet codons of the DNA language coded for which amino acids of the protein language. This began in the late 1950s and early 1960s. By the late 1960s, the genetic code had been solved. But another important discovery aided in the quest. That was the working out of the role of RNA in the process of cell heredity.

Before we turn to this, it must first be noted that paper chromatography also helped greatly in the working out of the structure of protein chains in the 1940s and 1950s. It also helped dispel the misunderstanding that was started by Levine's work, which was the belief that nucleic acid molecules were small alongside those of proteins. Let's look at how this came about.

THE USE OF PAPER CHROMATOGRAPHY

Although radioactive tracing and the centrifuge played a great role in the study of the chemistry of the genetic material in the late 1940s and early 1950s, one of the first real indications that DNA was the genetic material came with an application of paper chromatography beginning in the early 1940s. The chief pioneer in this area was the Austrian-born scientist Erwin Chargaff. The protein theory of the gene was still strong in the 1940s. That was because DNA could not be gotten in significant quantities from the cells of various organisms at the time. As we have seen, column chromatography was not well suited to analyzing small amounts of biochemicals. Also the centrifuge and radioactive tracing had been useful only in showing that DNA was the genetic material. They shed no light whatever on the chemical composition of the material in the 1940s. Paper chromatography had been shown to be useful in analyzing small amounts of certain proteins. By a slight modification of the method, Chargaff was able to work out the precise chemical composition of the small amounts of various DNAs from cells of different organisms. When classical chemical methods were used to analyze such small amounts of DNA, it had seemed that the modified four nucleotide theory of DNA structure had been established. Ordinary chemical procedures of the time could not precisely measure the numbers of moles of the four bases in very small amounts of DNA. So it seemed that all DNA molecules contained equal numbers of the four bases. But when Chargaff applied paper chromatography to different DNAs from many varied organisms, he found that the numbers of moles in one mole of different DNAs of the four ribose nucleotide substances were different for each. This meant that the DNA molecule could hold any number of base groups in any order along the polynucleotide chain. That is not all that Chargaff found out about DNA composition through paper chromatography, however. His precise work also showed that the amount (in moles) of adenine in any DNA was the same as the amount of thymine, while the amount of guanine was equal to the amount of cytosine. This meant that the number of adenine groups in any DNA molecule was equal to the number of thymine groups, while the number of guanine groups equaled the number of cytosine groups in any DNA molecule. By 1950 Chargaff's work had shown these regularities in composition of many different DNAs.

Chargaff had established two points. First, his work had finally dispelled the four-nucleotide theory of DNA composition that came down from Levine's work. This breakthrough showed that the DNA polynucleotide chain could be almost as complex in base composition and arrangement as the peptide chain of proteins was in amino acid composition and arrangement. It opened the door for the belief that the DNA molecule was indeed capable of enough structural variety to serve as the molecule of heredity. Secondly, this research established the restrictions in the base compositions of DNAs that are the basis of the A-T and C-G base pairing rules of the Watson-Crick model of DNA structure.

Watson and Crick made use of three areas of knowledge in the early 1950s to come up with their model of DNA structure. One of these was the results of many studies of DNA that had been done by means of X-ray scattering. The X-ray scattering patterns of many different DNAs revealed that the DNA molecule had some kind of helical structure. In addition, the base composition restrictions growing out of Chargaff's work could only mean that the DNA molecule had to be made of two helical polynucleotide chains wrapped around each other. There was no other way of explaining them. Finally Watson and Crick used the knowledge they had found available about the molecular structure of the four base groups to work out the A-T and C-G pairing rules of their model. But it must not be forgotten that Chargaff's findings were a necessity. Without them the Watson-Crick model could not have been postulated.

BONDING

What was the nature of the bonding between the A and T and the C and G groups? This question was not only important to an understanding of DNA structure, but also to an insight into knowledge of how more complex protein molecules are built. To answer the above question, we have to explore briefly how molecules form in terms of atomic structure.

How do simple molecules like those of water (H_2O) and table salt (NaCl) form? Let's first look at the water molecule (FIG. 9-2).

The hydrogen atom consists of a nucleus of one proton orbited by one electron. The electron is situated in the K shell. That shell can hold up to two electrons. The oxygen atom, on the other hand, generally has a nucleus of eight protons and eight neutrons, surrounded by eight orbital electrons. Two of the electrons go into the K shell, while six go into the L shell, which can hold up to eight electrons. The L shell in the oxygen atom is thus two electrons short of being full. So both the K shell of the hydrogen atom and the L shell of the oxygen atom are almost full. Chemists discovered that atoms have a strong tendency to fill their outer shells. So if two hydrogen atoms closely approach an oxygen atom, the electrons in the K shells of the hydrogen atoms and those in the K shell of the oxygen atom will be influenced by the nuclei. The two hydrogen atoms will share their K shell electron with the oxygen atom. Also the oxygen atom shares one of its L electrons with each hydrogen atom. In this way, each hydrogen atom effectively has a full outer shell of two electrons, and the

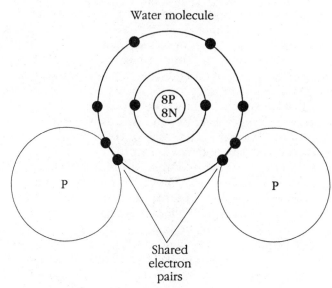

Water molecule

Shared
electron
pairs

9-2 The bonding of a water molecule consisting of one oxygen atom and two hydrogen atoms.

oxygen atom has a full outer shell of eight electrons. The oxygen atom shares a pair of electrons with each hydrogen atom, while each hydrogen atom shares a pair of electrons with it, and thus the water molecule is formed by a sharing of electrons between atoms to fill their outer shells. Each chemical bond between one of the hydrogen atoms and the oxygen atom is made up of a shared pair of electrons. Such a chemical bond is known as a *covalent bond*.

Now let's look at the smallest molecular unit of common salt, NaCl. The sodium atom consists of a nucleus holding 11 protons and about an equal number of neutrons, surrounded by 11 orbital electrons. Two of the 11 electrons go into the K shell while eight go into the L shell, so that the remaining electron goes into M shell in the sodium atom.

The maximum number of electrons an M shell can hold is 18; but in atoms in the weight range of the sodium atom and the chlorine atom, it can hold only up to eight electrons before becoming full. In the sodium atom the M shell holds one electron, making that shell far from full. The chlorine atom has a nucleus of 17 protons and about an equal number of neutrons surrounded by 17 orbital electrons. Two electrons go into the K shell, eight into the L shell, which leaves seven electrons for the M shell in the chlorine atom. So the chlorine atom has an outer shell that is almost full, missing just one electron. When the sodium atom meets the chlorine atom, the chlorine atom pulls the lone electron of the M shell of the sodium atom into its outer shell to make it full. But this leaves the sodium atom with one less electron than protons; that gives it a positive charge of one unit. On the other hand, the chlorine atom is left with an excess of one electron; this gives it a negative charge of one unit. Such charged atoms that have an excess or deficiency of electrons are known as *ions*. So when the smallest unit of common salt is formed, two oppositely charged ions are formed that attract each other strongly and stick together. The sodium atom has a full outer shell of eight electrons, an L shell, while the chlorine ion has a full outer shell of eight electrons also—an M shell in that case. This type of chemical bond, formed by the transfer of electrons from one atom to another, is known as an *ionic bond*. Salt, unlike water, does not really have molecules. Instead it consists of equal numbers of oppositely charged sodium and chlorine ions held together by electrical attraction.

From this discussion, there are two kinds of substances: *ionic* and *covalent*. Ionic substances are formed by electron transfer between atoms. Covalent substances are formed by electron sharing between atoms, and have molecules. You might get the impression that molecules of covalent substances are all alike in electrical properties because no electrons are transferred between atoms. You might think all molecules are electrically neutral. Nothing could be further from the truth. For one thing, what holds the molecules together in covalent substances? The force of gravity is much too weak between such small bodies to do the job. And if such molecules are electrically neutral, which, strictly speaking, they are, the electric force shouldn't be responsible for the attraction between them.

ELECTRIC FORCES AND POLAR MOLECULES

However, electrical forces are responsible for the attraction between the molecules of covalent substances. Molecules, like atoms, are made up of an electron cloud surrounding a number of positively charged nuclei. Although the molecules as a whole are electrically neutral, the atomic nuclei in one molecule attract the oppositely charged electron clouds of neighboring molecules. In light molecules this attraction is very weak. Therefore, the attractive forces between light molecules is usually very weak. But the larger molecules become, the stronger this attraction becomes. Small molecules like those of hydrogen, carbon monoxide, carbon dioxide, and ammonia, for example, are gases at room temperature. Larger molecules like those of ether CH_3OCH_3 and ethanol CH_3CH_2OH belong to liquids, while molecules larger than these belong to solid substances.

There is another kind of electrical force between molecules that is often stronger. Its existence is indicated by some simple observations. One of these was that water is a liquid at room temperature. Why was this so? All other substances with molecules of a similar size as those of water are gases. That can only mean that the attractive forces between water molecules are much stronger than would be expected on basis of the attraction between electron clouds and atomic nuclei of neighboring molecules. But why?

The nature of this second kind of electric force between molecules can be illustrated by considering a simple molecule like that of hydrogen, made of two identical atoms, and another kind of molecule like that of water. In the hydrogen molecule two electrons are shared between the two atoms, so that each has a full K shell. But each proton in the molecule attracts each electron with equal strength, since each of the protons has a single unit of positive charge. Thus, each electron spends equal amounts of time in each of the two atoms. And therefore each hydrogen atom in the molecule has no net positive charge. Now look at the water molecule. The angle between the two hydrogen atoms in the water molecule is 109.6 degrees. The reason for this is too technical to go into here. But the three atoms in the molecule are not lined up one after the other as the structural formula H-O-H tends to imply. Recall that each hydrogen atom and the oxygen atom share a pair of electrons. However, the proton of each hydrogen atom has only a single unit of positive charge, while the oxygen nucleus has eight units of positive charge. Therefore, the oxygen nucleus attracts the shared pairs of electrons much more strongly than the proton of each hydrogen atom does. As a result, the shared electrons spend more time in the oxygen atom than in the hydrogen atoms. The oxygen atom has a weak negative charge, while each hydrogen atom has a weak positive charge, although the electrons are not transferred totally to the oxygen atom, and the molecule as a whole is electrically neutral. Such a molecule is said to be a *polar molecule*. The oxygen end of the molecule has a weak negative charge, while the hydrogen end has a weak positive charge. The hydrogen molecule is an example of a *nonpolar molecule*.

The high electrical polarity of the water molecule explains why water is a liquid at room temperature. The oxygen atom of one water molecule attracts the oppositely charged hydrogen atoms of neighboring water molecules, making the attraction between such molecules quite strong. This attraction between a hydrogen atom of one water molecule and the oxygen atom of another is an example of what chemists call a *hydrogen bond*. Such bonding between atoms in different molecules is called *hydrogen bonding*. The electrons of the bonds between two atoms in many different molecules are often shared unequally between the two atoms. This gives certain parts of the molecule a positive charge and others a negative charge. The oppositely charged atoms of such molecules attract each other. Hydrogen bonding is said to exist between such molecules.

The phenomenon of hydrogen bonding helped Watson and Crick explain the A-T and C-G pairing rules between the two strands of the DNA molecule. The base groups of the two strands hold many hydrogen, oxygen, and nitrogen atoms, and the bonds between the hydrogen and oxygen atoms and between the hydrogen and nitrogen atoms are often polar in the base groups. The charged atoms of the groups are so arranged that an adenine group hydrogen bonds with a thymine group, while a guanine group hydrogen bonds with a cytosine group. But there are no hydrogen bonds between an adenine and cytosine group or between a guanine and a thymine group. Thus, the two strands (or chains) of the DNA molecule are held together by hydrogen bonds. There are two hydrogen bonds between each A and T group in the molecule and three hydrogen bonds between each guanine and cytosine group. We can see that the findings of atomic science shed much light on this important biochemical problem. This is a good illustration of the fact that all branches of science aid each other no matter how unrelated they might at first seem.

The concept of hydrogen bonding between atoms in different molecules also helped biochemists understand many of the details of the structure of the peptide chain. In fact, much headway was being made through the use of paper chromatography and the idea of the hydrogen bond in regard to this problem in the 1940s and 1950s. An English biochemist, Frederick Sanger, further developed the technique of paper chromatography introduced by Martin and Synge. His modification of the process was easily applied to the protein problem. The problem of protein structure consisted of three major parts: determination of the number of peptide chains in the protein molecule, determination of the number of different kinds of amino acid groups in each peptide chain of the molecule, and the determination of their arrangement in each chain.

THE PROBLEM OF PROTEIN STRUCTURE

To attack the three facets of the protein problem, Sanger developed an improved version of paper chromatography. It involved treating the protein being analyzed with a substance known as *dinitrofluorobenzene*, with the structural formula shown in FIG. 9-3. Sanger used the same substance when he applied chromatography to the protein problem in the late 1940s and early 1950s.

9-3 Structural formula of dinitrofluorobenzene.

Let's consider a protein molecule made of a single peptide chain to illustrate Sanger's method. Such a long peptide chain has the general formula shown in FIG. 9-4.

When the protein is treated with dinitrofluorobenzene, the chemical reacts with the amino acid at the end of the chain holding the -NH group and joins with it. This is shown in FIG. 9-5 for only the -NH_2 end of the chain. So the dinitrofluorobenzene treatment gives the modified peptide chain of FIG. 9-6.

Note that the dinitrofluorobenzene is attached to the amino acid group R_1 at the end of the chain. When the modified peptide chain is heated in acid solution it breaks into single amino acids and the modified amino acid of FIG. 9-7.

A small portion of the resulting mixture is placed on a piece of filter paper, and is then put through paper chromatography. In this way, the

9-4   Structural formula of a long peptide chain.

9-5   Partial illustration of reaction of amino acid and protein.

9-6   A modified peptide chain.

9-7   A modified amino acid.

amino acids in the mixture along with the above substance are separated. Each is then represented by a spot or blob on the filter paper after it is dried. But the blob made up of this substance is different from the rest. The substance has a bright yellow color. Therefore, its blob on the paper is also yellow in color. But the R group it contains, the R_1 group, represents the amino acid group at the end of the peptide chain. So by chemical means the substance in the yellow blob, or the amino acid at the end of the chain, can be established. This identifies one of the amino acids in the chain. The other blobs can now be analyzed to give the different kinds of amino acids in the peptide chain. The amounts of each amino acid in moles can also be determined from the blobs, and this tells the numbers of each kind of amino acid unit in the chain.

After the amino acid group at the end of the chain is identified in this way, another sample of the protein can be heated in mild acid solution. This has the effect of breaking the long chain into fragments. These can be separated by chromatography again. Consider one of them that has the formula of FIG. 9-8. When this smaller peptide chain is treated with dinitrofluorobenzene, it gives the chain of FIG. 9-9.

9-8 A small peptide chain.

9-9 Result of treatment of small peptide chain with dinitrofluorobenzene.

Using strong acid solution, all the peptide bonds between the amino acid units of a peptide chain can be broken to yield a mixture of simple amino acids. But the bond between the six-carbon ring group and the group -NHCHR$_4$ above is not broken by such treatment. So strong acid treatment breaks the smaller peptide into the substance of FIG. 9-10, and the amino acids represented by R_5, R_6, and R_7.

When the mixture is put through paper chromatography, four spots or blobs are obtained on the paper. The yellow colored one can be analyzed to figure out the nature of the amino acid side chain R_4. In this way the amino acid at the end of the short peptide chain can be determined. The other blobs can be chemically analyzed to tell what amino acids they represent. Now we know the amino acid R_4 at the end of the smaller frag-

9-10 Result of strong acid treatment of small peptide chain.

ment of the longer chain. Another small sample of the small peptide chain can now be obtained from the blob that represents it in the original chromatography treatment. When it is heated in mild acid solution (or if it is treated with special digestive enzymes in solution) it might break into the two smaller, simple peptides in FIG. 9-11.

9-11 Two small, simple peptides, A and B, that might result from heating of a peptide in a mild acid solution.

When the resulting mixture is put through paper chromatography, the two smaller peptides, each made of two amino acid units, separate as two blobs on the paper. Samples from these can be analyzed chemically. In this fashion, the first is found to hold the amino acids R_4 and R_5. Now it has been established by the dinitrofluorobenzene treatment that R_4 occurs at the end of the original four-amino acid fragment. Thus R_4 must come before R_5 in the first fragment above. But, in the original fragment of four amino acids, either R_6 or R_7 can come after R_5. How can it be determined which of them comes after R_5?

To answer this, another small sample of the original four-amino acid fragment from the larger peptide chain can be collected and again subjected to mild acid treatment. This time the fragment might break into the subfragments as shown in FIG. 9-12.

The mixture can be analyzed by paper chromatography again. In this way, the first fragment is found to be composed of the amino acids R_4, R_5, and R_6. The second is just the amino acid R_7. But the previous treatment established that R_4 comes before R_5. This can only mean that R_6 comes

A

$$H-N-C-C-N-C-C-N-C-C-O-H$$

with the structure showing H, H, O; H, H, O; H, H, O groups above and R_4, R_5, R_6 below.

B

$$H-N-C-C-O-H$$

with H, H, O above and R_7 below.

9-12 Two subfragments, A and B, resulting from mild acid treatment of a four-amino acid fragment.

after R_5 in the first fragment. But R_4 occurs at the end of the original four-amino acid fragment. Putting all these facts together, it would have to follow that the order of the amino acids in the original formula is:

$$R_4 - R_5 - R_6 - R_7$$

The amino acid arrangement of one of the fragments of the original peptide chain has been determined by paper chromatography and the dinitrofluorobenzene treatment. In the same way the amino acid arrangements of its other fragments can be determined. These arrangements are then compared to get clues about the amino acid arrangement in the original, long peptide chain. Then this long peptide is again treated with mild acid solution (or other digestive solutions) in order to break it up into different fragments. These are analyzed for their amino acid arrangements and compared with each other once again to give additional clues. After many such treatments of the original protein, the exact amino acid arrangement of its long peptide chain can be determined. This could be done by the 1950s, but it was a very time-consuming process.

Sanger's method can be applied to proteins of more than one peptide chain. If a protein of n peptide chains is treated with dinitrofluorobenzene, the substance reacts with the amino acid at the -NH end of each of its peptide chains in the same way as previously discussed to form a modified amino acid group at that end of each of the n peptide chains. Then, if the protein is heated in a strong acid solution, all the peptide bonds between amino acids are broken and the protein is separated into amino acids. But the group of FIG. 9-13 stays bound to each amino acid at the end of each chain. So, when the mixture is spread out into spots on filter paper through paper chromatography, n of the spots will have a bright yellow color. If n is 2, two yellow spots will be observed, and we know that the protein has two peptide chains. If n is 3, there will be three yellow spots, and the protein will have three peptide chains, and so on. The n spots can be analyzed for their amino acids, and various treatments can

9-13   A group that stays
bound to amino acids.
See text for details.

separate the n peptide chains in another sample of the protein from each other. These can then be subject to mild acid treatment or others and then put through paper chromatography to give their amino acid arrangements as in the example just given.

Several important proteins had been analyzed by Sanger's method by the early 1950s. He, himself, had worked out the exact structure of the enzyme beef insulin by 1953. Insulin is an important protein that regulates how an animal's body makes use of sugar in its life processes. Sanger used his method of chromatography to show that the insulin molecule had two connected peptide chains. They are held together by the atomic group -S-S- composed of two bound sulfur atoms, and the group is known as the *disulfide group*.

In the 1950s, it had been established that two kinds of bonds bind peptide chains together and give them their various shapes: the *disulfide linkage* and hydrogen bonding. The disulfide linkage comes about when two cysteine groups (in the same peptide chain or in different ones of a protein molecule) somehow get rid of the hydrogen atoms on the sulfur atoms on their side chains, so that the sulfur atoms have a free bond on them and can link up to each other. The structural formula of cysteine is shown in FIG. 9-14.

9-14   Structural formula of cysteine.

The formation of a disulfide linkage is given by FIG. 9-15. A disulfide linkage always binds two cysteine groups either in the same peptide chain, or in different ones in a protein molecule.

The two peptide chains of the beef insulin molecule are linked by two disulfide bonds. It was also found that one of its peptide chains branches back on itself through a disulfide linkage.

The hydrogen bond and the disulfide linkage are two of the most important bonds that give peptide chains their characteristic shapes. They

$$A-\overset{\overset{\displaystyle H}{|}}{\underset{\underset{\displaystyle H}{|}}{C}}-S-H \quad + \quad H-S-\overset{\overset{\displaystyle H}{|}}{\underset{\underset{\displaystyle H}{|}}{C}}-A \quad \longrightarrow$$

$$A-\overset{\overset{\displaystyle H}{|}}{\underset{\underset{\displaystyle H}{|}}{C}}-S-S-\overset{\overset{\displaystyle H}{|}}{\underset{\underset{\displaystyle H}{|}}{C}}-A \quad + \quad H-H$$

Disulfide
bridge

9-15 Formation of a disulfide linkage.

govern protein structure along with the ordinary covalent bonds between them.

Sanger's method of chromatography was only one that had been developed. Various others were invented and found useful besides. But his method is a good example. Today the whole process has been automated through many electrical and mechanical devices, so that the method is not as time-consuming as it had been in the 1940s and early 1950s. Yet the principle of the method is always as explained here.

RNA IN THE PROCESS OF HEREDITY

Better centrifuges had been developed in the 1940s. And, although RNA had been known to exist since the 1920s, nucleic acids could not be isolated from cells with any great efficiency until the centrifuge was developed. A high speed centrifuge not only separates cell organelles from each other but also separates groups of large biomolecules of different sizes from one another. The heavy organelles move toward the bottom of the tube, while the clearer part of the mixture holding many different biomolecules stays near the top of the tube. This liquid can be separated off the rest of the mixture in the tube, and spun at a greater speed in an ultracentrifuge. Then the heavier biomolecules move to the bottom of the tube, while the lighter ones stay near the top.

In this way, biochemists of the 1940s had uncovered some interesting facts about the DNA and RNA composition of different kinds of cells. For one thing, they found that the DNA content of cells from a given organism was constant; it did not vary in any of the cells of the organism. The same could not be said for the RNA content of the same cells. The RNA content of cells varied over a wide range. But one observation stood out: cells that were active in protein synthesis were much richer in RNA than cells that were not. Also, growing cells were found to contain more RNA than those that had reached maturity. Growing cells are much more active in building proteins than those that have stopped growing, since growing ones are building the materials they will need for their structure and function. So again, there seemed to be a relationship between RNA content and protein synthesis in the cell.

RNA and protein

These facts alone indicated that RNA was somehow involved in protein building in the cell. But how? Headway toward an answer started to be made in the 1950s. Scientists had then applied the centrifuge to separate nuclei from the liquid portion of samples of crushed cells. This left behind a liquid that held various lighter organelles of the cytoplasm. Among these, the mitochondria were the largest and most conspicuous. Interest in them increased. Scientists also applied the electron microscope to such mixtures of organelles from the cytoplasm, and in doing so, they saw that the mitochondria had an inner structure that has been described in chapter 2. Chemical analysis also showed that they held enzymes for the breakdown of fats and sugars into carbon dioxide and water, which showed that they were the powerhouses of the cell. Very little nucleic acid was found in the mitochondria. However, the electron microscope revealed that a large number of much smaller particles than the mitochondria existed in the mixture, and hence in the cytoplasm of the cell. These became known as *microsomes*, because of their small size, and when they were collected from the mixture by further use of the ultracentrifuge and chemically analyzed, they were found to be very rich in RNA. The microsomes became known as *ribosomes*, because of their high content of ribonucleic acid. They were also found to contain a lot of protein. They are a mixture of RNA and protein. Their high RNA content showed that they could not have been fragments of the cell membranes or the mitochondria. These structures held no RNA. They had to be independent organelles.

Zamecnik's experiment

These facts led to various speculations among biochemists in the 1950s. Could it be that the ribosomes were the sites of protein synthesis in the cytoplasm? To many, this seemed to be a sound hypothesis. But how could it be tested experimentally? One ingenious experiment that made use of radioactive tracing was done by Paul Zamecnik of the Massachusetts General around 1955. It made use of the radioactive isotope of carbon having a mass number of 14.

This isotope sends out electrons at a rate that can be detected by various particle detectors. Zamecnik made use of amino acids that contained carbon-14 instead of carbon-12 in his experiment. Thus the amino acids he used were radioactive. The experimental subjects in his research were rats. The liver in animals, is an organ that contains many cells active in protein synthesis. The radioactive amino acids were fed to the rats and passed into their systems. Zamecnik killed the rats at various times after the radioactive amino acids were injected into them, and their livers were taken out of them and crushed up. The various organelles were then examined for radioactivity, after they were separated from the mixture by methods previously discussed. In rats that were killed just after the radioactive amino acids were given to them, radioactivity appeared only in the ribosomes of the liver cells. In rats that were killed a little later, the radioactivity appeared in other cell organelles of the liver cells. In rats that were killed much later after the ingestion of the amino acids, the radioactivity of C-14 appeared in all cell organelles of the liver cells. This seemed to be clear proof that amino acids first move to the ribosomes after they enter the cell. There, they are assembled into peptide chains. Then these move to all parts of the cell. Zamecnik's work seemed to prove that the ribosomes are the sites of protein synthesis in the cell.

Zalokar's experiment

Another brilliant experiment, to give more direct evidence of the roles of the ribosomes and RNA in protein synthesis, was devised by the noted biochemist, Marko Zalokar, in 1960 at Yale University. He employed the technique of radioactive tracing with use of the centrifuge. The experiment was designed to find out two important pieces of information. One was to find out where free amino acids first go in the cell after being absorbed by it. This was the same problem that had been investigated by Zamecnik. The second was to find out where free base groups that comprise the RNA molecule go after being taken into the cell. How could such precursor RNA and proteins be followed in the cell? The answer was, once again, radioactive tracing. All amino acid and ribose nucleotide molecules contain hydrogen atoms. But no normal hydrogen atom is radioactive. However, the hydrogen isotope of mass number 3, or tritium, is very radioactive. In his experiment Zalokar used specially synthesized amino acids and ribose nucleotides that contained radioactive tritium instead of ordinary hydrogen of mass number 1. These can be given to the cells of an organism.

In the experiment, Zalokar made use of the amino acid leucine that contained radioactive hydrogen, or tritium. Besides, he used the RNA base uracil that was labeled with such hydrogen. The organism he used was a variety of the bread mold *Neurospora*. This mold consists of bundles of long, tube-like cells that point out from a given center in each bundle. These long, tubular shaped cells had two nice features that made them particularly useful in Zalokar's experiment. First, they contained not one but many cell nuclei. Thus they should yield a mixture rich in cell nuclei when they are spun in a centrifuge. Secondly, and really more important, the bundles or bunches of long cells could themselves act as centrifuge tubes that held all essential cell ingredients. Since the cells in each bundle point in the same direction from a point on the plant, each cell could act like a small centrifuge tube if a given bundle of them were spun rapidly on a small, but powerful, centrifuge. Thus, the various organelles in the long cells would be separated from each other to a great extent, and groups of a given type of organelle would take certain positions along the tube-like cells as they were spun. When a bundle of these plant cells was spun rapidly on a small centrifuge, Zalokar found that the nuclei gathered at some point of the long cell. The ribosomes gathered at a different point. He took note of both positions. Because of the special structure of the plant, he could spin each bundle of cells on a small centrifuge. That was very convenient. It would have been a lot more work to have to disrupt the cells and put the mixture into a centrifuge tube and spin it, as is usually done.

Zalokar could not feed the plant radioactive adenine, guanine, or cytosine because DNA contained these same bases. One of his aims was to find out where RNA is formed in the cell. It was known that DNA always stayed in the cell nucleus. But RNA was found in both the cytoplasm and nucleus. Zamecnik's experiment had shown that proteins are formed at the ribosomes. Zalokar's purpose was to further confirm this, and besides, to find out where RNA is formed in the cell and where it goes in the cell after it is synthesized. He wanted to follow RNA production through the cell. RNA contains uracil, and DNA does not. So radioactive uracil would have to be employed in this experiment.

He placed the Neurospora plants in a solution of uracil that contained a radioactive isotope. The long tubular cells absorbed the radioactive uracil. After about four minutes, he took some of the bundles of long cells from the plants and spun them in a small centrifuge. When he examined the cells with a radiation detector, he found that the radioactivity came from the point on each tubular cell where the nuclei of the cells gathered on being spun in the centrifuge. But he found no radioactivity in cells that had been exposed to the uracil solution for just one minute. This was clear evidence that RNA is synthesized in the cell nucleus. That seemed true because the nucleus was apparently the first place the radioactive uracil molecules went after being absorbed by the cells. In plants that had been exposed to the radioactive solution of uracil for eight to ten minutes, radioactivity was detected not only at the position of the nuclei in the centrifuged cells, but also at the position where the ribosomes had gathered. Here was clear evidence that RNA was made in the nucleus and then moves to the cytoplasm.

Next, Zalokar grew a group of the Neurospora plants in a medium containing radioactive leucine, a common amino acid in proteins. The radioactivity was first detected at the position of the ribosomes in the centrifuged cells. In plants grown for a much longer time in the solution, radioactivity was found in all parts of the centrifuged cells. These observations led to two conclusions. First, the leucine went to the ribosomes after being absorbed by the Neurospora cells. Secondly, it then went to all parts of the cell. The simplest assumption in these experiments was that the simple building blocks of the RNA and protein molecules first go to the place in the cell where they are assembled into the larger molecules of these biochemicals after they are absorbed by the cell. The larger RNA and protein molecules then migrate to other parts of the cell where they are needed. Since leucine went to the ribosomes after entering the cell, there was much direct evidence that protein chains are assembled at the ribosomes. Protein chains are found in all parts of the cell, both in the nucleus and cytoplasm. So it was not at all surprising that the radioactivity of leucine was later found in all parts of the plant cells.

Synthesis of RNA

The following picture was suggested by experiments like these: RNA is made in the cell nucleus. The only other nucleic acid in the cell nucleus is DNA, the genetic material. The DNA molecules therefore had to somehow control the making of other molecules in the cell. So, it seemed at this time—the late 1950s and early 1960s—that DNA directed the making of RNA, which then migrated to the cytoplasm to form ribosomes. The ribosomes somehow governed the assembly of peptide chains from the amino acids that built them. The process was not quite that simple. But these findings did strongly suggest that DNA made RNA, which made protein. The overall scheme of cell heredity expressed by

$$DNA \rightarrow RNA \rightarrow Protein$$

was seen as a reasonable way to describe the main theme of cell heredity. So, at that time, RNA came to be seen as the middle man that carried the genetic message from the nucleus to the cytoplasm where protein is made.

At first it was reasonable to suppose that RNA from the nucleus took the form of ribosomes that directly made the proteins the cell needs. The ribosomes were first thought to be the makers of protein, and were not thought of as merely the sites where the process takes place. Instead it was envisioned that the bases in the RNA of the ribosomes determined the type of protein in a given ribosome would make. It was believed that the base arrangement of each kind of gene in the DNA of the nucleus somehow determined the base arrangement of the RNA in a given kind of ribosome, which, in turn, determined the amino acid arrangement of the peptide chain coded for by the gene. This hypothesis at first seemed obvious and self-evident as well as simple.

Even in the late 1950s, certain facts were noted that cast doubt on this simple scheme. For example, experiments making use of radioactive tracing that were conducted in 1956 showed that a new kind of RNA, made when a virus invades a cell, existed that had a base composition resembling that of the invading virus rather than that of the DNA of the cell. The new kind of RNA appeared in the cell shortly after a single virus entered the cell. This could be explained in two ways according to the above prevailing hypothesis. Either the viral DNA quickly made ribosomes of its own that made the protein characteristic of the virus, or the invading virus held ribosomes already made when it entered the cell. The first of the two possibilities was clearly unlikely. The infection process caused by the virus in the cell was known to be very rapid. Now ribosomes were shown to be rather complex molecules of both RNA and protein. It was not very likely that so many of them could be built in such a short time. The second of these possibilities also seemed quite farfetched. Many viruses were barely bigger than ribosomes themselves. They could hardly contain enough ribosomes to account for the speed of the infection process.

Then there was the problem of the rate of protein production by the cell. Sometimes cells stepped up the rate of production of a given protein when it became necessary to do so. But various tracing experiments tended to show that the cell contained no more ribosomes when the rate of protein production was high than when it was low.

Then, finally, there was the matter of the base composition of ribosomal RNA. This had been determined for cells of many different organisms by the late 1950s. While the DNA composition varied from organism to organism, the RNA composition of the ribosomes often did not match the DNA composition and was found to be fairly constant over a wide range of organisms.

The Brenner-Jacob-Meselson experiment

For these reasons, doubts about the theory that the ribosomes actually made proteins began to arise. But the first experimental proof that the ribosomes were not the makers of proteins came in 1961. In that year the scientists S. Brenner, Meselson, and F. Jacob devised an experiment that conclusively showed that the ribosomes could not serve this purpose. It is known as the Brenner-Jacob-Meselson experiment. In the experiment the investigators grew a colony of *E coli* cells in a medium containing the heavy nonradioactive isotopes of carbon and nitrogen—C-13 and N-15. The medium was not radioactive. The cells had been growing long enough in this medium so that all their organelles contained these heavy, nonradioactive isotopes. The *E coli* cells were then transferred to another medium that contained only the two lighter nonradioactive isotopes of carbon and nitrogen—C-12 and N-14—and the radioactive form of phosphorus, P-32. As soon as the cells were transferred to this medium, they were infected with T4 viruses. The assumption behind the procedure of the experiment were simple. The *E coli* cells, at the time they had been put into the new radioactive medium, had cell organelles and biomolecules that held C-13 and N-15 atoms only, since they had been grown for a long time in the nonradioactive medium. But they began to absorb C-12, N-14, and radioactive P-32 atoms when they were put into the new medium. Now if the DNA cores of the T4 viruses that entered the cells built ribosomes to make their protein shells, they should have used the C-12 and N-14 atoms that the cells were now absorbing to do so. All the ribosomes that had existed in the cells before infection held the atoms of the old medium—C-13 and N-15. The phosphorus atoms in them would also have to be the nonradioactive kind. But if new ribosomes were made by the viral DNA cores after they invaded the cells in the new medium, these ribosomes should have held C-12, N-14, and P-32 atoms. They should have also been radioactive because of the radioactive phosphorus they contained.

After the viruses infected the *E coli* cells, the cells where disrupted and spun in a density-gradient centrifuge. The investigators then examined the centrifuge tube for radioactivity. If the viruses made new ribosomes in the cells, the radioactivity should have shown up at a point in the tube where the density of its liquid was equal to that of ribosomes holding N-14, C-12, and P-32. But a strange thing was found. The label of the radioactive phosphorus was found at a position in the tube where the density of the liquid was close to that of the old ribosomes that had existed in the cells prior to infection by the viruses which held the heavier isotopes C-13 and N-15. None of the ribosomes that gave off the P-32 radioactivity were found to contain any C-12 or N-14 of the new medium. Why was the radioactivity of P-32 associated with the cell's ribosomes? There was no way this finding could be explained if the viruses made ribosomes of their own when they invaded the cells. To do so they would have had to use the N-14 and C-12 of the new growing medium.

During the years before this experiment, there had been some interesting speculations, however. Some experts envisioned that the DNA of the nucleus made a single stranded type of RNA that had a base sequence complementary to the base sequence of the gene it represented. These single stranded RNA molecules would migrate to the ribosomes. The ribosomes would somehow interact with this new kind of RNA and the free amino acids of the cytoplasm to help make the protein chains needed by the cell. This new single stranded RNA was called *messenger RNA*, since it would carry the genetic message from the cell nucleus to the ribosomes. The ribosomes were somehow involved in protein synthesis. Experiment had clearly shown that. The single stranded molecule of messenger RNA would originate in the nucleus and would move to the ribosomes. The ribosomes somehow aided in the putting together of the peptide chains the codons of the specific messenger RNAs coded for.

The Brenner-Jacob-Meselson experiment conclusively showed two things. One of these was that the DNA of the T4 virus did not make new ribosomes. The fact that the ribosomes they utilized contained C-13 and N-15 meant clearly that they were using the cell's ribosomes. Secondly, the fact that the ribosomes were associated with a substance containing radioactive phosphorus (P-32) of the new growing medium meant that the viral DNA was taking RNA bases made of this isotope of phosphorus, and making a new RNA that associated with the ribosomes of the *E coli* cells, while making proteins needed for the shells of the virus. Further chemical analysis managed to isolate small traces of this RNA. Its base sequences proved to be complementary to those of the DNA of the T4 virus.

So the Brenner-Jacob-Meselson experiments had shown that the ribosomes do not directly make protein. They only aid in the process in some way. It also proved the existence of a new kind of RNA: messenger RNA. This new RNA served the role originally assigned to the ribosomes. The messenger RNA is what was found to be made directly by the DNA of the genes in the nucleus. It carried the genetic message from the nucleus to the ribosomes. It was found that the strands of messenger RNA held the genetic message in the cytoplasm and not the ribosomes. The ribosomes only assisted in translating the genetic message in the messenger RNA into the protein language. Each strand of messenger RNA also carried the same genetic message as the gene that made it in the nucleus. This message was converted into a given peptide chain that the gene coded for at the ribosomes. So it came to be seen that messenger RNA was the link between the cytoplasm and the nucleus as far as the genetic message is concerned. Radioactive tracing experiments had clearly shown that protein chains are put together at the ribosomes.

When a virus invaded a cell, its DNA core made its own messenger RNAs that used the ribosomes to make the peptide chains of the viruses' protein shell. The radioactive phosphorus (P-32) found associated with the *E-coli* ribosomes in the Brenner-Jacob-Meselson experiment came from such messenger RNA. The messenger RNA joined with the cell's ribosomes in protein synthesis.

How do the ribosomes aid in the making of protein chains? How do the free amino acids in the cytoplasm recognize the genetic message held in the strands of messenger RNA? These are the questions that yet had to be answered in the early 1960s.

TRANSFER RNA

Early in the 1950s, Francis Crick came up with an idea that said that the liquid portions of the cytoplasm might contain relatively small RNA molecules that each held an *anticodon* of each possible codon on a messenger RNA molecule. An anticodon of a given codon is simply one that is complementary to the given codon. If the codon is AGC, its anticodon is TCG. If the codon is AAT, its anticodon is TTA, and so on. At the time Crick put forth this hypothesis, the ribosome was still thought to be the maker of protein. Therefore, in Crick's hypothesis, each kind of ribosome would contain the RNA codons for the making of a given protein chain. Each of these codons would stand for one amino acid. He envisioned that the free amino acid molecules would be joined chemically to the small RNA molecules. The amino acid molecule would be joined to the small RNA molecule at one of the RNA molecule's ends. At the other end of the small RNA molecule, there would be an anticodon of a codon of that particular amino acid. This anticodon, being complementary to the RNA codon for the particular amino acid, would bind to the codon of the amino acid on the ribosome. Therefore, through the small RNA molecule holding the anticodon of the amino acid, the amino acid would become bound to the ribosome. There would be other such small RNA molecules holding the RNA anticodons of other amino acids that could be chemically joined to their molecules at the other end of the small RNAs. Thus after the first amino acid of peptide chain coded for by a ribosome was attached to the ribosome through one of these small RNA molecules, Crick envisioned that another such small RNA molecule would join to the next codon on the ribosome through its anticodon. This second RNA would hold the anticodon of the next amino acid to be incorporated into the peptide chain of the ribosome, and would be joined chemically to that amino acid. The two amino acids would thus be lined up before each other on the ribosome. And the same would happen for the other amino acids coded for by the ribosome. So all the amino acids for the peptide chain of the ribosome would be lined up on the ribosome through the small RNA molecules. They could then be joined to each other by enzyme-controlled chemical reactions. This would form the peptide chain coded for by the ribosome. The peptide chain and the small RNA molecules would then be freed from the ribosome. Then the small RNAs would again join to other free amino acid molecules, and the process could repeat.

This hypothesis for the making of peptide chains by ribosomes seemed sound at the time. But there was a deficiency. As was already known, ribosomes were very small organelles. Their RNA chains would have to fold compactly to be contained in them. So how could all the base triplets of ribosomal RNA be expected to bind to the anticodons on the smaller RNA molecules? How could all the amino acids of a large peptide chain be lined up on such a small particle as a ribosome? Maybe a ribosome underwent many structural modifications as it made a peptide chain. In such modifications, maybe only a few RNA codons of the gene of the ribosome were exposed at a time. Then the small RNA molecules postulated by Crick could attach themselves to the ribosome a few at a time, and the amino acids would be joined to the ribosome a few at a time. The ribosome would continually undergo such structural modifications as it made the peptide chain, exposing only a few codons at a time. But this process would be unduly complicated. There were no clear mechanisms for such intricate changes in the structure of the ribosome.

Then there were also the previously mentioned difficulties with the theory that the ribosomes were the makers of peptide chains. These were well known at the time.

In 1957, the American biochemist Mahlon B. Hoagland discovered a type of RNA in the cell that fitted the description given by Crick earlier. It had small molecules—about 70 nucleotides on the average. It was also soluble in the cell fluids. He called it *soluble RNA* for this reason. By the early 1960s, the properties of this RNA had been further studied. Its molecule held an average of 70 nucleotides. The molecule of this RNA had complex shapes. Some of them, for example, are shaped like a cloverleaf.

By that time, biochemists had come to the conclusion that there must be something in Crick's earlier suggestion about such RNA in the cytoplasm of the cell. Yet the details of protein manufacture in the cell could not be as simple as the ribosome theory had pictured them. After all, messenger RNA had been discovered, too. It was this RNA that carried the genetic message from the nucleus to the cytoplasm.

At this point a possible scheme of protein synthesis began to occur to many biochemists. What if the anticodons of the amino acids on the soluble RNA molecules attached themselves to the codons on the messenger RNA molecules instead of to the ribosomes? Then the smaller soluble RNA molecules would carry the amino acid units to the messenger RNA molecules in the cytoplasm instead of to the ribosomes. But it was also an experimental fact that protein synthesis occurred at the ribosomes. So what real function did the ribosomes serve in the process?

The idea occurred to many experts that the ribosomes made protein synthesis easier for the cell by acting as a binding site for both the messenger RNA molecule and the soluble RNA molecule. A ribosome moves along a strand of messenger RNA. As it does so, it stays bound to the strand of messenger RNA. But as it moves along, various soluble RNA molecules (carrying their corresponding amino acid units) become bound to the ribosome at its other side, one after another. So, when the ribosome begins its movement along the messenger RNA at one end of the strand, it comes to the first RNA codon of the strand that stands for some amino acid. At the same time, a molecule of soluble RNA, holding the amino acid unit of the first RNA codon at one end and the anticodon of the amino acid at the other, becomes bound to the ribosome also. Then the anticodon on the soluble RNA molecule binds to the codon on the messenger RNA molecule and is thus held on the messenger RNA strand. The ribosome then moves on to the next codon on the messenger RNA strand. Then a soluble RNA molecule, having the amino acid and anticodon of this codon, is joined to the ribosome also. Again, the second codon on the messenger RNA strand and the anticodon on the soluble RNA molecule join, and the ribosome moves on to the third codon of the messenger RNA strand. So the first two amino acids of the peptide chain coded for by the messenger RNA strand are now joined to the strand. By complex chemical changes, a peptide bond is formed between the two amino acid units. The two soluble RNA molecules are released from both the messenger RNA strand and the amino acid units, and return to the liquids of the cytoplasm and are used again. In the same way, the third amino acid unit of the peptide chain is joined to the third messenger RNA codon and to the first two amino acid units of the peptide chain being formed. The same process continues as the ribosome moves down the messenger RNA strand. When it gets to the end of the strand, the peptide chain is completed, and is released from the messenger RNA strand and the ribosome. In theory, as a ribosome moves along a messenger RNA strand, it helps put together the peptide chain of the particular messenger RNA. The function of the ribosome is to act as a surface on which the messenger RNA codons are joined to the corresponding soluble RNA molecules, which hold the amino acids coded for by the codons. The ribosome functions something like a catalyst in this way. When a ribosome reaches the end of a messenger RNA strand, it goes on to another messenger RNA strand. The small molecules of soluble RNA serve to transfer the free amino acid units in the cytoplasm to the messenger RNA strands where they are assembled into peptide chains. For this reason, this kind of RNA became known as *transfer RNA*.

As the 1960s progressed, more and more experimental facts spoke for the truth of this theory of protein synthesis. The theory also gave the most productive clues as to how to break the genetic code. What RNA codons code for which amino acid units? If this question could be answered, we would immediately know which DNA codons code for which amino acid units. After all, the strands of messenger RNA are complementary to the DNA ones they are copied from in the nucleus. So the two codes are equivalent. Either the DNA code or the RNA code give the same genetic information. Often the RNA codons for the amino acid units are the ones given in many sources. These are just as informative as the DNA ones, except that the base group U in one code replaces T in the other.

In the early 1960s, the time of these developments, biochemists and biologists began to prepare what are called *cell-free systems*. The centrifuge aided them greatly in this feat. A cell-free system is a liquid medium gotten from disrupted cells that hold no cell organelles, except ribosomes, but contains all the other important substances in the cytoplasm of the cell. These include amino acids, messenger RNA, transfer RNA, and others that supply and transfer energy in cell processes. Such media also held ribosomes.

Around the same time, an enzyme was isolated from cells that could bind free ribose nucleotides together. It could be added to solutions of the four ribose nucleotides, and when it was, it would bind the nucleotides into single strands of RNA of various base compositions. So when the enzyme was added to a solution of adenylic acid, it formed long chains of RNA holding only adenine units. This long RNA chain could be represented as:

AAAAAAAAAA. . . . . .

If the enzyme was added to a solution of urridylic acid, an RNA made of only uracil units was made. It could be represented by:

UUUUUUUUUU. . . . .

These were simple messenger RNAs. Various cell-free systems were prepared that were freed of most of the messenger RNAs they already held. Then a simple messenger RNA like UUUUUUUU. . . . was added to them. If the prevailing theory of protein synthesis was correct, the ribosomes and transfer RNAs of the system should use this simple messenger RNA to form a simple peptide chain, made up solely of the amino acid unit coded for by the RNA triplet UUU. After this simple messenger was added to such a system, peptide chains holding only the amino acid unit phenylalanine were isolated from the system. This showed that the RNA codon UUU coded for the amino acid phenylalanine. When a simple messenger RNA like AAAAAAAA . . . . . was added to such a cell-free system, long peptide chains holding only lysine units were isolated from the system. This could only mean that the RNA codon AAA coded for the amino acid lysine. In the same way a simple messenger RNA like CCCCCCCC . . . . . gave peptide chains made only of proline from such systems. The RNA codon CCC coded for proline. In the same way, the RNA codon GGG was shown to code for the amino acid glycine. Thus, the amino acids for four of the 64 codons were gotten. However, that was only a start. Much remained to be done.

In the case of simple messenger RNAs holding only one base unit, the problem was simple. But what if man-made RNAs holding more than one base unit were used? Then the problem was much more complex. Consider a solution of uridylic acid and guanylic acid. If the messenger RNA making enzyme is added to this solution, the enzyme will form simple RNA chains made of both base units—U and G—so that there would be more than one kind of codon in the RNA chains formed. Two of them would be UUU and GGG. But then there would also be those that hold two G units and one U unit. These would be GGU, GUG, and UGG. Also, there would be those that held two U units and one G unit. They would be UUG, UGU, and GUU. So there would be eight possible codons in the messenger RNA formed, and the frequency of the codons along the RNA chain could be predicted from the proportions of the two ribose nucleotides used in making the RNA chain. When these chains are added to a cell-free system, peptide chains containing several amino acids are formed. Yet it was difficult to tell which amino acid corresponded to each of the codons in the synthetic RNA. It was clearly known that UUU and GGG stood for phenylalanine and glycine, but it was harder to tell which of the amino acids in the new peptide chains the remaining six codons stood for in this approach. But the method did yield strong clues.

A more powerful method of determining which amino acids the RNA codons coded for was devised by Philip Leder and Marshall W. Nirenberg of the National Institutes of Health in 1964. They made use of very simple messenger RNAs having molecules that consisted of single codons. They then added these codon molecules to a cell-free system containing ribosomes, and all the transfer RNA molecules joined their respective amino acids, as well as other essential components for protein synthesis. They then added each simple three-base RNA to a number of such systems. They knew the base arrangement of the codon they had added to each cell-free system. But the experiment also made use of radioactive tracing. Each cell-free system contained all the amino acids joined their respective transfer RNAs. But in each such system just one of the amino acids was made radioactive and contained a radioactive isotope. It was a different amino acid in each of the cell-free systems. Leder and Nirenberg added each of the 64 simple three-base RNAs to each of the cell-free systems. The codon that each three-base RNA represented, and the transfer RNA containing its anticodon and amino acid, would bind to the ribosomes of the cell-free system to form relatively large particles, composed of the transfer RNA molecules, ribosomes and the amino acids. Then, the experimenters passed the solution composing each cell-free system through a nitrocellulose filter. Nitrocellulose is a porous substance with many microscopic holes. When each cell-free system was passed through such a filter, all molecules and ribosomes would pass through the small holes, except for the large particles composed of the three-base RNA molecule, its transfer RNA molecule, and a ribosome. These would be left behind in the filter, since they were too large to pass through the holes in the filter.

If the filter showed no radioactivity after all the liquid passed through it, Leder and Nirenberg knew that the radioactive amino acid in that particular cell-free system was not coded for by the RNA codon they had added to that system. They did the same for each of the cell-free systems, and the first time radioactivity showed up in the filter they knew that the three-base codon they had added to that system had attached itself to the transfer RNA of the radioactive amino acid the system contained. They also knew which amino acid was radioactive in that system, and the base arrangement of the codon they had added to it. They then knew which amino acid the RNA codon they added to the system coded for.

In this way, Leder and Nirenberg determined the RNA codons of many amino acids. Through this method and the others just discussed, the genetic code was solved by 1967. Knowing the messenger RNA codon of each amino acid, the DNA codon was known for the amino acid. The RNA codons of the amino acids are given in TABLE 9-1. The codon AUG serves to start the reading of the genetic message on a messenger RNA molecule. The codons UAA, UGA, and UAG serve to stop the reading of a messenger RNA strand. They code for no amino acids.

Table 9-1 The Genetic Code

Glycine	GGU, GGC, GGA, GGG
Leucine	CUU, CUC, CUA, CUG, UUA, UUG
Isoleucine	AUU, AUC, AUA
Lysine	AAA, AAG
Cysteine	UGU, UGC
Glutamine	CAA, CAG
Glutamic acid	GAA, GAG
Alanine	GCU, GCC, CGA, GCG
Arginine	CGU, CGC, CGA, CGG, AGA, AGG
Asparagine	AAU, AAC
Aspartic acid	GAU, GAC
Phenylalanine	UUU, UUC
Methionine	AUG
Proline	CCU, CCC, CCA, CCG
Serine	UCU, UCC, UCA, UCG, AGU, AGC
Valine	GUU, GUC, GUA, GUG
Threonine	ACU, ACC, ACA, ACG
Tyrosine	UAU, UAC
Tryptophan	UGG
Histidine	CAU, CAC

CELL HEREDITY PROCESS SUMMARY

The scheme of cell heredity can be summed up as follows: When a given peptide chain is needed by the cell, an enzyme molecule in the nucleus—messenger RNA polymerase—moves along a stretch of DNA comprising the gene of the peptide chain. As this molecule moves over the DNA double helix, it temporarily unwinds its chains to expose the bases on them to free ribose nucleotides floating around in the nucleus. These nucleotides are then attached and bound to their complementary bases on one of the DNA chains of the gene. They are bound to them through the pairing rules A-U and C-G. As they are bound to one of the DNA chains, the ribose nucleotides are joined together through various chemical processes, a few at a time, as the molecule of messenger RNA polymerase moves along the DNA double helix. After the enzyme molecule passes a given portion of the DNA double helix, the two DNA chains wind up again and free a small portion of the messenger RNA strand of the gene. When the enzyme molecule reaches the end of the DNA chain of the gene, a molecule of messenger RNA has been assembled from the ribose nucleotides in the liquids of the nucleus. This process is called *transcription*. In the process, the codons making up the DNA language of a gene are converted to the RNA codons of the RNA language. The messenger RNA molecule of the gene then moves to the cytoplasm. There, with the aid of the transfer RNAs and the ribosomes, it makes the peptide chain coded for by the gene in the nucleus. This process is called *translation*. The RNA language is translated into the protein language at this stage.

The RNA of the ribosomes is also copied from DNA in the nucleus. There are certain genes there for this purpose. After ribosomal RNA is copied from them, it gathers in the structure known as the *nucleolus* of the nucleus before it goes to the cytoplasm in the form of ribosomes. Other genes give the transfer RNA molecules.

That only one chain of the DNA molecule is copied in making messenger RNA has been shown by experiments. It became possible in the 1970s to get significant amounts of the DNA of a given gene. When such DNA is heated, it melts. Such melting was found to separate the two strands of the DNA molecule into single strands. When a messenger RNA of the gene of the DNA is added to the mixture of the two chains, it is found to bind only to one of the strands. That means that the messenger RNA is complementary to only one of the DNA chains; it can only have been transcribed from that DNA chain.

By 1970, the genetic code was solved. Many important developments have to be omitted in a book of this scope. There are other good popular books that go into some of the other developments in this field, and the interested reader is advised to consult them if interested.

Chapter **10**

Genetic engineering

The nature of the gene was a mystery at the time Mendel speculated about it in the nineteenth century. But just a century later the genetic code had been cracked. What is the real significance of such a breakthrough?

One consequence of this new knowledge has been many practical applications of modern genetics. Not all of these could be touched upon in even a large book. But it would not be fair if a book of this nature did not give some time to the question of the promise modern genetics holds for the future of humankind.

MANUFACTURED PROTEINS

Along these lines, the technique of *genetic engineering* holds the most promise. This branch of genetic technology has developed during the 1970s and 1980s, and now offers many possible benefits that were not even thought of when the genetic material was discovered in the 1940s.

The essence of genetic engineering is simple in principle. It utilizes the one-celled organisms that were so helpful in earlier genetic research. *E coli* cells are one example. The ribosomes of such cells will use any messenger RNA supplied to them to help make the protein coded for by the messenger RNA. The RNA need not come from the cell's DNA. The DNA of a virus makes its own messenger RNA that the host cell's ribosomes use to make the protein needed by the virus. But the DNA introduced into the cell need not be that of a virus. It could be the DNA of any gene from another organism. The essence of genetic engineering was the discovery of a way to insert the DNA of a given gene into a host cell such as an *E coli* cell. If the DNA of such a gene could be inserted into the cell, it would produce its own messenger RNA in the cell just as the DNA of viruses do. This messenger RNA would use the ribosomes of the cell to produce the protein chains coded for by the gene.

Headway to performing this feat had been made by the 1980s. Certain enzymes had been discovered that could break the DNA chains in a cell only at certain places in the chains. These are known as *restriction enzymes*. The DNA in human body cells contains many different genes. Significant amounts of human DNA could be obtained by 1980. Also certain restriction enzymes were found that could chop this DNA into much smaller chains, each about the size of an average Mendelian gene. So the probability is high that some of the DNA molecules among the smaller ones obtained are those of a given gene sought after by the genetic engineer.

What does the genetic engineer do with the smaller DNA chains? Another variety of enzymes had been discovered called *ligases*. These have the ability to join two or more of the DNA fragments into circular pieces of DNA known as *plasmids*. Some of the plasmids obtained by their use might contain the gene or genes sought after by the genetic engineer. Ways have been found to insert various plasmids into bacterial cells.

Such plasmids have been found to have two interesting properties. Many of them can function and produce messenger RNAs in other cells. Also, those that can function in the cells are reproduced when the cells divide and are passed to the daughter cells from generation to generation. So if the genetic engineer can find various plasmids that hold certain genes that function in bacterial cells, he can produce large quantities of the protein of the gene. Once a few such cells that contain the plasmids of the given gene are isolated from a herd of cells into which the plasimds have been inserted at random, these few can be grown in large numbers in a culture. This culture can produce the protein of the gene in large amounts.

Many important protein substances of value in medicine, industry, and research have been obtained in quantity through genetic engineering. An important protein that diabetic individuals need to regulate their sugar metabolism, *insulin*, is now being made through genetic engineering. Insulin made in this way has a high degree of purity, and is totally human insulin. Until recently, diabetics had to use insulin obtained from animals. Another protein made in quantity by genetic engineering in recent years is the hormone known as HGH—*human growth hormone*. Humans who lack this hormone remain small in stature, or have a disease known as dwarfism. HGH regulates growth in normal individuals, and people who do not produce it do not grow. These two examples show that genetic engineering holds great promise for the curing of various diseases.

That is only one area where the quickly developing fields holds potential for the future. The chemicals that industry produces are often made through the use of inorganic catalysts. Though such catalysts speed up the production of these chemicals, the reactions still require a lot of energy input. But more energy input means more expense for the industry. One potential genetic engineering holds for industry is the possible mass production of enzyme-like catalysts. These enzyme-like catalysts would be protein in nature and would speed up the reactions that produce a given chemical. There would be one important difference here. Recall that enzymes make possible the chemical reactions of life at a very low temperature. Thus not much energy has to be supplied to get the reaction going. With inorganic catalysts of industry at present, much energy must be supplied just to start a reaction that produces an important chemical. Much money could be saved by means of these new catalysts.

The promise that genetic engineering holds for the future could take up an entire book in itself. Yet the field has already shown that our understanding of the genetic code has made possible many marvelous developments that could improve our standard of living for the better. We owe it all, however, to Mendel's first efforts to unravel the mystery of heredity.

MAPPING THE HUMAN GENOME

The Human Genome Project, sponsored by the National Academy of Sciences, began in early 1986. Since then, scientists have successfully mapped the relative positions of several hundred genetic markers on all 46 human chromosomes. Such a map, known as the human genome, carries the complete set of instructions for making a human being. It might someday pinpoint all 100,000 or so genes that contain the human complement of hereditary material. The genetic map will help researchers identify and devise prenatal tests for a number of inherited diseases that are the result of two or more genes. Maladies such as heart disease, cancer, mental illness, and other common disorders might have multiple genetic roots. With this map, it is possible to unravel the complicated inheritance patterns of these diseases.

The physical map of the human genome is like a line of bottles. Each bottle contains DNA fragments of about 40,000 bases long, whose position on the genome is accurately known. Producing individual fragments like these, known as *cosmids*, has been an important part in the biologists' tool kit for some time. The challenge is to blanket the entire 30 billion bases of the human genome with overlapping cosmids so that there are no gaps in the map. That's a tough task, especially for some parts of the genome which have long stretches of repeated sequences. The project is expected to take many years and cost billions of dollars. But the potential rewards are tremendous.

Although it is still some years in the future, medical doctors envision a time when they will practice so-called *gene therapy*, in which they actually enter the cells to repair their genetic machinery. They do this by a process of gene splicing, in which genes are chemically removed from the DNA of one organism, joined to the DNA of another organism, and inserted into a bacterium, which is grown to produce large quantities of the gene's protein. The first protein made by these methods was insulin, the hormone essential for metabolizing sugar. Since it was first introduced in 1982, it has proved a blessing to diabetics who are allergic to traditional sources of insulin.

Gene-splicing proved to have the potential for treating certain other genetic diseases. One method of doing this is by using viruses to transport normal versions of a gene to cells in which the gene is missing or defective. The lack of such a functioning gene prevents the production of some vital protein. In *hemophiliacs*, for example, portions of the gene responsible for a blood-clotting enzyme are missing. By inserting normal versions of the gene into the patient's bone marrow cells, the missing enzyme might be produced in sufficient quantities to cure the disease. This, however, is only one of more than 4000 inherited diseases. By mapping the human genome, medical scientists will be able to find the responsible gene's location on one of the 23 chromosome pairs. Once pinpointing the exact location of the gene, this could lead to a screening test to identify genetic disorders in individuals. This is especially important for couples who have family histories of genetic defects and want to have children.

Alzheimer's disease is an affliction in which the victim suffers a gradual loss of memory and ends up in a vegetative state. The disease is linked to the buildup in the brain of a certain peptide, which is produced by the genes that code either for the peptide itself or for the enzyme that breaks down the peptide. Researchers have successfully developed a screening test for families with a history of the disease. Finding a cure for the disease will require a tedious search through the chromosomes and examine bits of DNA to locate the genes that are responsible. This could take many years. In the meantime, patients who test positive for the disease must agonize over how they will conduct their lives. The future might seem so horrifying that some people might even refuse to take the test and thus irresponsibly pass the defective gene on to their offspring.

MORAL AND ETHICAL QUESTIONS

It would seem that manipulating an embryo to correct a defective gene and thereby preventing a birth defect or the chance of inheriting a debilitating disease would be a good thing. However, a more ominous use of the technology might be to select certain favorable traits, producing a super-race of human beings. We have seen this before during Hitler's regime in Germany in the 1930s and 40s, when certain individuals who fit the ideal model were selected to become baby factories. People might look upon the technology as a genetic equivalent of cosmetic surgery, in which parents could impart to their offspring traits that are deemed favorable to the cultural convention of the time.

Fetuses that are found to have genetic disorders could be aborted to save families the agony of bringing into the world a child who suffers and will probably die young in life. Even worse, abortions might be conducted if the fetus has the wrong sex or a slight abnormality. The parents might feel that this child is going to be costly to raise, and opt to abort and try for another. Prospective spouses might want a genetic profile of each other so that complications such as these do not arise. Society might even demand that couples be stopped from passing genetic deformities on to their children. Those deemed genetically inferior might be sterilized, ending forever their line of heredity.

In other areas, genetic screening might be used to select individuals for certain tasks for which they are genetically better suited. For example, prospective workers, who might be exposed to hazardous working conditions such as chemicals or radiation, might be tested for a genetic disposition toward cancer. Insurance companies could refuse to provide coverage for individuals who are certain to develop inherited diseases later in life. Airline pilots might find themselves grounded because a gene linked to alcoholism was found even though the pilots themselves do not use alcohol. This could explode into genetic discrimination, whereby only the best physical specimens are allowed the complete freedoms of society.

On the farm, genetic engineering provides a panacea that appears to be the solution to many of the world's food problems. Certain diseases or drought resistant strains of crops have been developed to improve agricultural output. Some plants have been genetically engineered to produce larger fruits. Even farm animals are genetically engineered to produce cows that give more milk and pigs that grow to prodigious size through the use of human growth hormones. When the embryos from sheep and a goat were fused in a test tube and implanted into a goat's womb, a *geep* was the result. The animal is a sterile hybrid with a goat's head on a sheep's body. It sparked a scientific and ethical debate whether such bizarre creatures should be patented so that corporations could benefit financially by tinkering with nature.

Moral and ethical issues must be faced now, before genetic engineering gets out of hand. A broad social debate is needed, dealing with the dangers as well as the benefits. But perhaps the most obvious danger is that genetic screening will provide one more excuse to divide up human beings into superior and inferior races. The genetic age could bring about new discrimination and anxiety. Just as with the development of the atomic bomb, the genetic genie cannot be put back into its bottle and will be with us from now on. Scientific advances often far outpace institutions designed to implement their safe use. Hopefully, despite the danger, the knowledge will be used wisely for the benefit of all people.

Glossary

alleles Different forms that a given genetic trait can take.

amino acid One of the twenty simpler substances out of which the complex proteins comprising the living cell are composed which have the general structural formula $NH_2 - CH - COOH$.

atom The smallest particle of a chemical element having the properties of the element.

autoradiography Type of radioactive tracing making use of imprints on photographic emulsions.

bacteria A large category of one-celled organisms that often invade larger organisms.

bacteriophages A group of viruses that invade a bacterial cell and reproduce inside it.

biomolecules Large organic molecules that comprise the materials of the living cell like those of proteins, nucleic acids, carbohydrates, and fats.

centrifuge A device making use of rapid rotary motion to separate different organelles in the protoplasmic materials obtained from living cells.

chromosomes Long threadlike bodies that appear in the nucleus of a cell before cell division that contain the genes.

condon A triplet of DNA bases that either codes for an amino acid or serves to start or stop the process of transcription.

coenzymes Simpler substances forming part of the molecular structure of some enzymes that are essential to the functioning of the enzyme.

conservative method Postulated method of DNA replication in which the two polynucleotide chains of the original DNA molecule stay together in one of the daughter DNA molecules.

crossover The exchange of parts between chromosomes that pair up in the process of meiosis.

cytoplasm Part of the cell outside the nucleus where most of the cell's metabolism takes place.

DNA Deoxyribonucleic acid or the genetic material.

double helix Picture of the DNA molecule that portrays it as composed of two helical polynucleotide chains wrapped around each other.

drosophila melanogaster A small fruit fly used extensively in early chromosome mapping experiments.

E coli Type of bacterial cell that lives and multiplies, particularly in the human intestine, that found much use in modern genetic experimentation.

electron A light negatively charged particle that comprises part of the atom.

enzymes A class of complex proteins that help promote the chemical reactions of the living cell.

fatty acid A group of complex organic acids that combine with glycerol to form various fats.

formula A shorthand notation used by chemists to indicate the atomic makeup of molecules.

fructose One of the three simple six-carbon sugars that form part of the molecular structure of more complex carbohydrates found in the cell.

galactose Another of the three simple six-carbon sugars that may comprise part of the structure of more complex carbohydrates.

gamete A reproductive cell such as an egg cell or sperm cell of parent organisms.

gene The functional unit of heredity that governs the passage of a given trait from parent to offspring.

genetic code The sequence (or ordering) of bases along the DNA chain of a gene that leads to the production of a given peptide chain (or chains) governed by the gene.

glucose The most important of the three simple six-carbon sugars comprising part of more complex carbohydrates, that is responsible for supplying the cell with energy.

hemoglobin A conjugated protein in red blood cells that carries oxygen to body cells.

heredity The passing of traits from parent organisms to offspring.

hydrogen bond A weak attraction between two atoms in a large molecule that does not arise from electron sharing, which is strong enough to hold the molecule in a given configuration.

infection The process by which viruses invade cells and reproduce inside them.

meiosis The process of germ cell formation in which the number of chromosomes is reduced to half the normal number found in somatic cells of the organism.

mitosis Ordinary division of cells that does not reduce the number of chromosomes.

messenger RNA A single stranded RNA that carries the genetic message from the DNA of the nucleus to the ribosomes of the cell.

mitochondria Oval shaped organelles of the cell, in which sugars and fats are broken down to produce energy to keep the cell operating.

molecular genetics Branch of genetics that seeks to explain the process of heredity on the molecular level of cell chemistry.

mutation A chemical modification of a gene that changes the trait the gene controls, and that can be brought about by X-rays or certain chemical substances.

nonsense codons Condons that do not code for an amino acid.

nucleic acids Substances that are the genetic materials like DNA, or those such as messenger RNA or transfer RNA that carry out the genetic process of transcription or translation.

nucleolus A small organelle in the cell nucleus in which ribosomal RNA is synthesized.

nucleus A small organelle in the cell that contains the genetic material and controls the chemical activities of the cell.

one gene-one enzyme hypothesis The idea that one gene might control the production of one enzyme needed by the cell.

peptide chain A long chain of amino acid units that makes up part of or all of a protein molecule.

protein The chief and most complex substances of the living cell that comprises its enzymes and its structural materials.

replication The process in which a chromosome or DNA molecule makes a duplicate or copy of itself using cell materials.

ribosomes Very small oval organelles in the cell, composed of RNA and protein, that are the sites of protein synthesis in the cell.

RNA Ribonucleic acid.

RNA polymerase An enzyme in the cell nucleus that temporarily separates the DNA chains and exposes the bases on them to ribose nucleotides in genetic transcription.

semiconservative method Postulated method of DNA replications, in which one of the two new polynucleotide chains exists in each of the two daughter DNA molecules.

transcription The process by which a strand of messenger RNA is constructed from the strand of DNA of a given gene which takes place in the cell nucleus.

tRNA Transfer RNA.

ultracentrifuge A very high speed centrifuge capable of separating large molecules of different kinds from one another.

viruses Large nucleoprotein molecules that can invade a cell and reproduce inside.

X-ray diffraction   The scattering of X-rays by crystals of substances that gives important information about the molecular structure of the substances.

zygote   The first living cell of an organism formed by the union of a gamete from each parent.

Bibliography

American Foundation for Continuing Education. *The Mystery of Matter*. Oxford University, 1965.

Asimov, Issac. *A Short History of Biology*. The Natural History Press, Garden City, NY, 1964.

_____. *Asimov's Guide to Science*. Basic Books, 3rd Ed., NY, 1972.

_____. *The Chemicals of Life*. New American Library, NY, 1954.

_____. *The Genetic Code*. New American Library, NY, 1962.

Browder, Leon W. *Developmental Biology*. Saunders College, Philadelphia, PA, 1980.

Carey, John and Joan Hamilton. "The Genetic Age." *Business Week*. May 28, 1990: 68–83.

Cohn, Norman S. *Elements of Cytology*. Harcourt, Brace, and World, Inc., 2nd Ed., NY, 1969.

Giese, Arthur C. *Cell Physiology*. W. B. Saunders Company, 5th Ed., Philadelphia, PA, 1979.

Gribbon, John. *In Search of The Double Helix*. McGraw-Hill Book Company, NY, 1985.

Jenkins, John B. *Genetics*. Houghton Mifflin Company, Boston, MA, 1975.

Keeton, William T. *Biological Science*. W. W. Norton & Company, NY, 1967.

Lehninger, Albert L. *Short Course in Biochemistry*. Worth Publishers, Inc., NY, 1973.

Magner, Lois N. *A History of the Life Sciences*. Marcel Dekker, Inc. NY, 1979.

Merritt, Jim. "Design of Life," *Modern Maturity*. Vol. 32. June-July 1989: 43–47.

Nouikoff, Alex B. and Eric Holtzman. *Cells and Organelles*. Holt, Rinehart, and Wilson, Inc., NY, 1970.

Readings from *Scientific America. Facets of Genetics*. W. H. Freeman and Company, San Francisco, CA, 1970.

Rogers, Michael. *Biohazard*. Avon Books, NY, 1979.

Sylvester, Edward J. and Lynn C. Klotz. *The Gene Age*. Charles Scribner Sons, NY, 1983.

Index